dies in
nadian
ography

Etudes sur
la géograph...
du Canada

e Prairie
vinces

Les provinces
de la Prairie

Edited by/Sous
la direction de
P. J. Smith

ed for the 22nd International Geographical Congress
à l'occasion du 22e Congrès international de géographie
al 1972

sity of Toronto Press

© University of Toronto Press 1972
Toronto and Buffalo

ISBN 0-8020-1921-8 (Cloth)
ISBN 0-8020-6161-3 (Paper)
Microfiche ISBN 0-8020-0258-7

Printed in Canada

Contents

Foreword

The publication of the series, 'Studies in Canadian Geography,' by the organizers of the 22nd International Geographical Congress, introduces to the international community of geographers a new perspective of the regional entities which form this vast country. These studies should contribute to a better understanding, among scholars, students, and the people of Canada, of the geography of their land.

Geographical works embracing the whole of Canada, few in number until recently, have become more numerous during the last few years. This series is original in its purpose of re-evaluating the regional geography of Canada. In the hope of discovering the dynamic trends and the processes responsible for them, the editors and authors of these volumes have sought to interpret the main characteristics and unique attributes of the various regions, rather than follow a strictly inventorial approach.

It is a pleasant duty for me to thank all who have contributed to the preparation of the volume on the prairie provinces. A special thanks is due to: Mr R.I.K. Davidson of the University of Toronto Press; Mr Geoffrey Lester who guided the Cartography Laboratory of the Department of Geography, University of Alberta in preparing all the illustrations; the Canadian Association of Geographers for its financial support; and the Executive of the Organizing Committee of the 22nd International Geographical Congress. Finally I wish to thank Professor P.J. Smith, chairman of the Department of Geography at the University of Alberta, for having accepted the editorship of this volume.

LOUIS TROTIER
Chairman
Publications Committee

Avant-propos

Par la publication de cette série d'« Etudes sur la géographie du Canada », les organisateurs du 22e Congrès international de géographie ont voulu profiter de l'occasion qui leur était donnée de présenter à la communauté internationale des géographes une perspective nouvelle des grands ensembles régionaux qui composent cet immense pays. Ils espèrent que ces études contribueront aussi à mieux faire comprendre la géographie de leur pays aux Canadiens eux-mêmes, scientifiques, étudiants ou autres.

Les travaux d'ensemble sur la géographie du Canada, peu nombreux jusqu'à récemment, se sont multipliés au cours des dernières années. L'originalité de cette série provient surtout d'un effort de renouvellement de la géographie régionale du Canada. Les rédacteurs et les auteurs de ces ouvrages ont cherché moins à inventorier leur région qu'à en interpréter les traits majeurs et les plus originaux, dans l'espoir de découvrir les tendances de leur évolution.

C'est pour moi un agréable devoir de remercier et de féliciter tous ceux qui ont contribué d'une manière ou d'une autre à la réalisation de cet ouvrage sur les provinces de la prairie. Il convient de mentionner les membres du Comité d'organisation du 22e Congrès international de géographie; M. R.I.K. Davidson, des Presses de l'Université de Toronto; l'Association canadienne des géographes; le département de géographie de l'Université de l'Alberta, à Edmonton, dont le Laboratoire de cartographie a préparé toutes les illustrations de cet ouvrage sous la direction habile et dévouée de M. Geoffrey Lester. Je remercie également le professeur P.J. Smith, directeur de ce département, d'avoir accepté d'assumer la direction de cet ouvrage.

LOUIS TROTIER
Président du
Comité des publications

Preface

This series of original essays is intended to provide a contemporary geographical view of a region which has been too easily misunderstood, not just in descriptive writing but, with greater damage, in the minds of the manipulators of economic and political power in Canada. The view from the interior is well conveyed by the following extract from an editorial in the *Edmonton Journal* (5 February 1972). It refers to a specific grievance but exemplifies a much broader concern and frustration – the frustration of being misunderstood and undervalued.

High freight rates are a fact of life in this country, particularly in the west ... It wouldn't be so bad if they were just high. But some of them are outright discriminatory, favoring one region over another. How many people in Central and Eastern Canada realize that it is cheaper to ship some commodities to Edmonton via detour through Vancouver than to Edmonton directly? It is also cheaper to ship Alberta-grown oil seeds east and west than the refined product, thus stifling development of a provincial food processing industry.

Every Western Canadian ... is familiar with the inequalities of the Canadian railways' freight rate structure. Farmer, manufacturer and consumer alike all share the burden. But how often are they listened to in the board rooms of the two railways' headquarters, in the corridors of the federal government's regulatory authorities? Tokenism and promises to look into things are all they ever get.

The theme of misunderstanding comes through in almost all the following chapters. It begins, indeed, with the physical base since the regional name is itself a misnomer, and has led to a grossly oversimplified view of the environment of the prairie provinces. The physical base is actually characterized by great diversity through space and great variability through time. Hence Professor Richards' suggestion that the common title, 'the prairies,' should be replaced by 'the western interior.'

Misunderstanding of the physical environment has reacted on the human use or misuse of the region in many ways, and this provides a further theme for several authors. The weaknesses of institutional attitudes toward the size of homestead grants, for example, is a topic in Chapters 2 and 3, and has an obvious bearing on the rural depopulation trend which is discussed in Chapter 5. The prairie provinces make a vital contribution to the national economy but, as Chapter 4 makes plain, it is very difficult to break out of the neo-colonial role of a primary resource exporter. Deliberate attempts to manipulate the physical, economic, and social environments of the provinces have been an inevitable outcome, and the theme of increasing government intervention is one that is stressed in all chapters from the second to the last.

To a very large extent the problems of the prairie provinces, and their present geographic character, are a reflection of four basic facts: the huge size of the region, its small population, its comparative isolation, and the recency of its effective settlement. In a sense, these are also national problems, bearing particularly on Canada's relations with the United States and Western Europe, but within Canada they bear mostly heavily on the prairie provinces. None of the authors seems to hold out much hope for any appreciable change in the foreseeable future.

The arrangement of chapters is intended to provide a progression from the physical base through man's impact in a historical context to the principal components of the economic geography of the region. It then proceeds to the population patterns and urban forms which have been produced by this complex interplay of physical, social, political, and economic forces. The book concludes with a Retrospect and Prospect, which serves largely to draw together the major threads of the preceding chapters. As a synthesis, though, it could equally well serve as an introduction to the region, and perhaps it should be both the first and the last chapter to be read.

The chapters were all individually conceived and written. The general framework was of the loosest kind to allow maximum flexibility to the authors to stress those themes which they felt to be most critical. The result, then, does not stand as a comprehensive evaluation of the prairie provinces, and is not offered in that vein. All the authors, too, felt the pressures of space very keenly. All could have written at much greater length, and I have had to exercise my editorial duty with much greater firmness than I would have wished. In particular, I have tried to eliminate the most serious duplication of content, though when the duplication was necessary to the arguments of more than one author I have allowed it to stand. One valuable consequence is that the authors have unwittingly demonstrated, through the divergences of their own views, the conflicts that bedevil the region. In his evaluation of the resource base, for instance, Professor Laycock views the economic future with much greater equanimity than either Professor Barr or Professor Proudfoot. Is he allowing himself to be fooled by the apparent abundance and variety of these resources? or is the optimism of a native Albertan showing through? or have Professors Barr and Proudfoot taken too jaundiced a view of the ability of the prairie economy to survive the vagaries of the international marketplace and the infirmities of the power brokers of central Canada? How 'difficult' is the prairie environment for agriculture? To several authors it is hazardous and unreliable, but Professor Laycock again takes an optimistic view, despite his close familiarity with the limitations and variability of the physical environment. And how important is it to have more people? Professors Barr and Richards stress the inhibiting force of a small domestic market, but Professor Nelson brings forward the notion of strictly controlled growth. Is a larger population an economic necessity or an ecologic disaster? Or is the question academic, since Professors Weir and Richards both cast doubts on the region's long-term growth potential? These and many other questions are raised in the chapters that follow.

University of Alberta   P.J.S.
February 1972

1 The Diversity of the Physical Landscape

ARLEIGH H. LAYCOCK

The prairie provinces of Alberta, Saskatchewan, and Manitoba occupy the southern interior of western Canada. Their combined area of almost 2,000,000 square kilometres (757,985 square miles) is approximately one-fifth of Canada's total, and the sheer size of the region explains much of its physical diversity.

The diversity of physical and cultural landscapes has not been stressed enough in most descriptions of the prairie provinces. The simple picture of a great, flat expanse of natural grassland is obviously inadequate, but so are descriptions which focus upon the broad zonal divisions which characterize the region in the minds of most geographers. The zonal divisions certainly exist, at a broad level of generalization, but use of them may obscure as much as it reveals of the character of the prairie environment. The geomorphic patterns, for instance, are complex, and their interaction with the other physical elements ensures an almost infinite assemblage of landscape units. Similarly, the climatic and moisture patterns vary critically and often dramatically over short periods and distances. Their local interaction with the other elements of the environment at any given time may be so far removed from the impression conveyed by long-term averages and zonal patterns as to make them unrecognizable. The diversity of the prairie landscape, then, is both spatial and temporal, and the interactions of these dimensions account for the many local variations in 'natural' patterns, and for much of the variety in human activity within the region.

Physiography and Geology
The structure and lithology of the prairie provinces are diverse, yet large areas have relatively uniform patterns and the major divisions can be described briefly (Figures 1.1 and 1.2).

The Canadian Shield is the core of most of the continent. In Precambrian time it extended well beyond its present limits but in succeeding times it was at least partly flooded by seas which advanced over it and

1.1
Relief and Drainage

Elevation

500	1000	2000	3000	4000	6000	over 6000 Feet
152.4	304.8	609.6	914.4	1219.2	1828.8	over 1828.8 Meters

⎯⎯ Major drainage divides

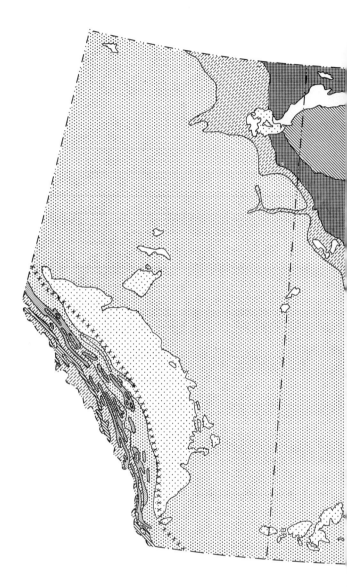

x x x x x **Foothills and Plains Margir**

1.2
Geological Features
(Source: Geological Survey of Canada)

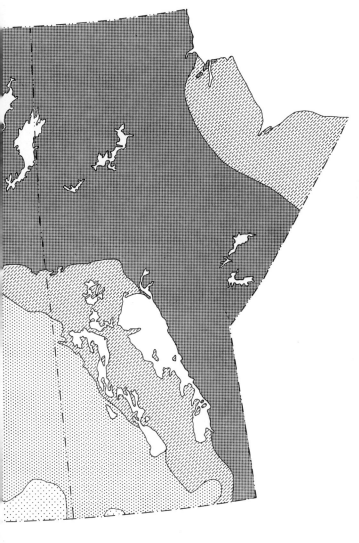

Tertiary

Cretaceous

Mesozoic
in Mountains and Foothills

Palaeozoic

Late Proterozoic

Archaean
and Early Proterozoic

later retreated. The accumulated sediments have been largely removed, but Palaeozoic and Mesozoic remnants form the Hudson Bay lowlands and much of the great depth of sedimentary formations of the interior plains. The Shield has been a relatively stable mass since Precambrian time, but its earlier history was very complex, resulting in widespread mineralization which has been exposed through peneplanation. The relief of the area now is subdued, with maximum elevations of 600 metres. The most striking surface feature is the disorganized drainage, which is the product of glacial scouring and deposition. There are hundreds of thousands of angular rock basin lakes connected by streams with irregular gradients and tortuous courses. These separate knobby uplands in which local relief is usually under 100 metres. The only extensive exception is on old sedimentary rocks south of Lake Athabasca, where the plain is gently rolling with more rounded lakes and less angular stream courses.

The interior plains cover approximately half the prairie provinces and include almost all the oecumene. They are underlain by sedimentary rocks which dip gently toward the southwest, away from the Shield, except where there are more recent continental deposits which dip toward the northeast. The older ones are exposed in bands of varying width west of the Shield margin, but deep Upper Cretaceous deposits generally prevail at the surface. In the south and west they, in turn, are overlain by more recent continental deposits. Minerals which are associated with these sedimentary formations include bituminous and lignite coal, oil and gas (in Devonian reefs, in stratigraphic and structural traps, and in oil-bearing sands), potash, salt, and various construction materials (e.g., gypsum).

In contrast to the geological formations, the plains surface slopes gradually from an elevation of 1200 metres in the southwest to below 220 metres near the Shield margins. Stream gradients are gentle; yet the plains are not completely flat, open, and featureless. There are numerous upland areas, such as the Cypress and Swan hills in Alberta and Riding Mountain in Manitoba, which rise between 300 and 800 metres above the adjacent lowlands. Lesser relief is associated with numerous other erosional remnants, and with morainic hills and the valleys of pro-glacial and post-glacial streams.

Descriptions of the plains make frequent reference to three steps or levels. The first is the Manitoba lowland below the Manitoba escarpment, the second extends to the foot of the Missouri Coteau, and the third to the foothills. This is a recognizable sequence in parts of the south but it is less so farther north, and a misleading impression of simplicity has been created.

The mountain and foothill areas have rocks ranging in age from Pre-cambrian to Recent. The oldest, largely metamorphosed sedimentaries, compose the major fault blocks in the south and are the bases of many others near the continental divide. The massive limestone formations in most mountain blocks are Palaeozoic in age. They were deposited in a great geosyncline that received more than 5500 metres of Cambrian sediments in the Kicking Horse Pass area, with an equally large deposition occurring through to the Mississippian. Early Mesozoic sedimentation, and volcanic deposits from mountain building to the west, added shallow water and continental deposits of sandstone, shales, and coal. Major uplift occurred in the Upper Cretaceous, and the subsequent erosion contributed large amounts of sandstone and shale to the fluctuating shallow seas covering the interior plains. The Laramide Orogeny in the early Tertiary produced major folding, faulting and uplift, and the parallel arrangement of fault blocks and fault-associated ridges which prevails in the mountain and foothill zones today. Down-faulting has also preserved deposits of coking coal and other minerals (e.g., limestone and gypsum) and created structural traps for oil and gas. Local relief is intense in the mountain region, with the ridge crests often 2000 metres above the floors of the intervening valleys, despite the modifying effects of Pleistocene glaciation. The foothill belt to the east has lower elevations, less relief, and more rounded forms, but its broken terrain is still as much a barrier to settlement as the more dramatic mountain landscape or the rock and water wilderness of the Shield.

Landforms and Surface Materials
At the beginning of the Pleistocene the climate of the prairie provinces apparently changed from warm and dry to cool and moist. The pedi-plains that had developed along the mountain front and around the smaller uplands experienced extensive valley erosion and local alluviation. Lakes and marshes were rare and residually weathered materials covered most surfaces. Four major glaciations then occurred, with ice from Kee-watin and Hudson Bay centres spreading across the Shield and into the plains. Almost simultaneously, cordilleran glaciers moved down into the foothills and plains. It is likely that the last glaciation was the most exten-sive, but relatively little is known about the earlier ones, or about the short-term interstades during which major ice-front recession occurred. At the time of maximum glaciation, however, the cordilleran and conti-nental ice flows coalesced in the Alberta foothills and plains margin. Only small summit areas, as in the Cypress and Porcupine hills and isolated mountain nunataks, were not covered (Figure 1.3).

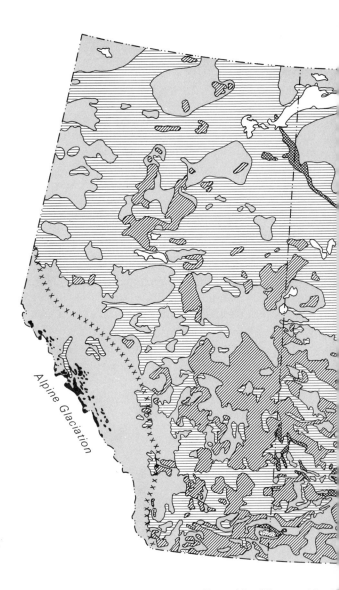

Zone of Cordilleran and Conti
x x x x x   Ice Coalescence

1.3
Surficial Features
(Source: Geological Survey of Canada)

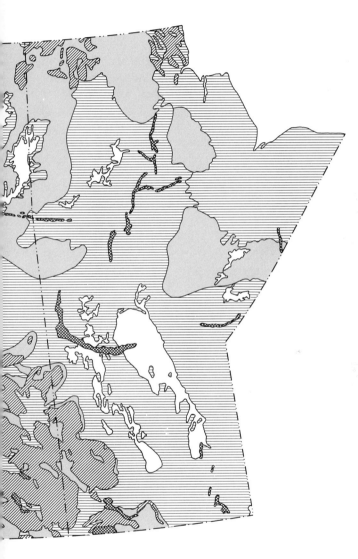

Existing glaciers

Unglaciated

Ground moraine

Lacustrine

Hummocky moraine

End moraine

The Shield experienced major ice scour and much of its weathered surficial material was transported to the south and west, along with numerous large erratics. Many valleys were filled with drift and knobs were rounded. Local relief was probably reduced in many areas but ice plucking in the more fractured rock resulted in millions of angular depressions. This and the blockage of old channels have resulted in the present disturbed drainage pattern. Post-glacial changes have been minor and there has been little weathering of surfaces or incision of stream channels. Most rivers have numerous rapids and waterfalls separated by lakes, marshes, and other stretches with relatively flat gradient. Most soils are stony, of variable depth, and poorly drained.

The interior plains received very large amounts of drift from the Shield, yet most deposits are of relatively local origin. These include limestone erratics from the older sedimentary rock areas and finer materials from the softer, younger shales and sandstones. Many of the materials have been reworked several times, and drift deposits are as much as 300 metres in depth over some pre-glacial valleys. A very important feature of deglaciation was that the general slope of the land was toward the wasting continental ice mass. Meltwater and land surface flow were ponded against the ice front at successively lower levels across the plains. Major pro-glacial lakes were formed, and their lacustrine deposits are extensive and widespread. Many other surfaces were bevelled by wave action and depression filling. The lacustrine and lacustro-till deposits are excellent for agriculture because of the flat to undulating topography, the fine moisture-retentive quality of the soils, and the easily improved drainage condition. The successive spillway channels incised through divides and roughly paralleling the receding ice front are striking features of the plains landscape. Many are occupied, at least locally, by present streams, and they have an effect of deflecting flow southward for short distances from a northeasterly course. Many terminate in delta deposits formed on the sides of pro-glacial lakes. Most of the aeolian sand plains, some with dunes of up to 30 metres in local relief, have developed on delta and outwash deposits. Very few are still active.

The apparently 'dead-ice' condition of the melting ice contributed to the development of distinctive hummocky moraines in many areas. Many marginal moraines with outwash and push features and numerous kettles are also present, often on the flanks of pre-glacial erosional remnant hills and cuestas. The local relief and numerous sloughs in moraine areas are disadvantages but cultivation is not uncommon because the till is often relatively stone-free and erosion problems are not as great as in more humid regions. Drumlins and eskers are not numerous in the south, but many are present in upland portions of the northern plains and the Shield. Numerous outwash and spillway terrace deposits are excellent sources of sands and gravels.

Most landscapes have not been changed greatly by post-glacial erosion. Aeolian deposits are widespread and a thin veneer of the loess has been added to most surfaces. Weathering is contributing to a breakdown of the softer shale and sandstone materials but local water surpluses are too small to transport much material. Nonethless, the larger rivers rising in the mountain and foothill areas have incised valleys from 30 to 250 metres into parts of the plain and their local tributaries are also very active. Much of the region might now be identified as a young stream-eroded plain. Many deltas and floodplains have been deposited on the flanks of lakes and on lacustrine plains. Those of the Peace, Athabasca, Saskatchewan, and Red rivers are the most extensive.

The cordilleran glaciers in the west were largely of local origin although significant amounts of ice moved eastward across the divide, forming low-level 'through-valley' passes. Alpine glaciation was intense at all levels and most of the characteristic erosional and depositional features are present. Alpine glaciers and icefields persist today in the higher mountain areas, but they are the product of approximate balances between winter snowfall and summer melting and are not remnants of Pleistocene ice. The largest is the Columbia Icefield, which feeds a number of prominent glaciers. Post-glacial changes include the erosion of canyons by tributary streams on the flanks of major valleys; the development of floodplains where valley gradients were shallow; major slide development where unstable formations were left after valley deepening by moving ice; and the irregular clothing of slopes by vegetative cover. All have added to scenic diversity, and the recent focus upon recreation in parts of the region is not surprising. Other wildland uses including watershed and wildlife management, forestry, commercial grazing, and mining are similarly affected by landform features.

Climate

The climate of the prairie provinces is the variable product of many controls. Latitude contributes to the mild summers and cold winters, and the continental interior location accentuates this seasonal temperature range and limits precipitation. These are accentuated in turn by the mountain barriers to the west, which inhibit the flow of mild air and moisture from the Pacific and ease the meridional flow of cold air from the north, particularly in winter, and of warm air from the south, particularly in summer. The relative proximity of the Pacific and Hudson

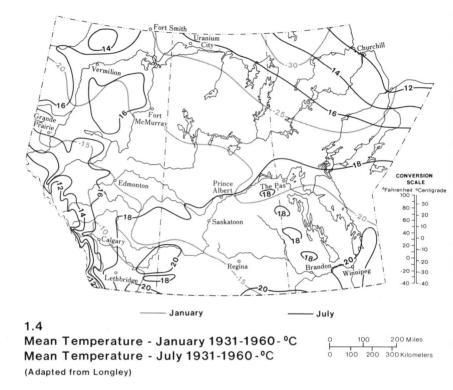

——— January   ——— July

1.4
Mean Temperature - January 1931-1960 - °C
Mean Temperature - July 1931-1960 - °C
(Adapted from Longley)

Bay, affected respectively by their mild and cold currents, is partial explanation for downward temperature gradients from southwest to northeast. These are modified in summer, except near Hudson Bay, by elevation decreases in the same direction. These controls are fixed, yet they
provide the bases for the air mass and pressure contrasts that produce
the storms and air mass successions that so characterize the climate of
the region.

If average temperature and precipitation data are used, the climate of
the prairie provinces could be classified as Dfc (subarctic) in the north
and Dfb (humid continental, cool summer) in the south with H (highland)
climates in the southwest. With slightly different periods of record, a
small area of BSk (mid-latitude steppe) could be added in the southern
plains. If annual data are used, the BSk climates occupy much of the
south in the drier years, though they are variable in extent and intensity.
Similarly, in the colder years the Dfc boundary might shift far southward,
while in the warmer years it would be north of the lower Peace River
valley. These changes greatly affect agriculture and most other aspects of
life in the provinces.

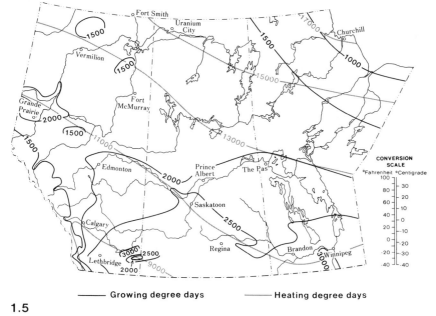

——————— Growing degree days   ——————— Heating degree days

1.5

Growing Degree Days - Annual - 1921-1950 - °F above 42°F (5.6°C)
Heating Degree Days - Annual -1921-1950 - °F below 65°F (18.3°C)

(Source: Chapman and Brown (1966) and Thomas, Climatological Atlas of Canada)

The average temperatures in January grade sharply downward from the southwest (Figure 1.4), reflecting the dominance of cold, polar continental air in the northeast. The southwest is often under the same air mass, but average temperatures are moderated by invasions of warm, polar marine masses from the Pacific. With the addition of the heat of condensation from heavy snowfall in the mountains and adiabatic warming during their descent down the eastern slope, they produce the warm 'chinook' (foehn) winds that so modify winter weather for varying periods. Temperatures are cold in continental air masses, but sensible temperatures are moderated slightly by the low wind, low humidity, and high sunshine factors. Winter blizzards with temperatures of −30°c to −40°c and high winds also occur in the plains, and extreme wind-chill temperatures of −60°c may be experienced. An index of annual heating requirements may be established if 65°F (36.1°C) is assumed to be a desired temperature level and heating is assumed to be proportional to the number of degree-days below this level. Approximately twice as much fuel is needed in the northeast as in the southwest (Figure 1.5).

The July temperature gradient is shallow, and maximum levels occur in a broader band across the southern plains (Figure 1.4). Solar radiation is the major control, and the longer northern days and lower elevations of the northern and eastern plains help to compensate for the weak air mass contrasts. Proximity to Hudson Bay results in steeper gradients in the Shield and Hudson Bay lowlands. Some of the highest temperatures (sometimes over 43°C) are recorded in the drier southern areas. Air mass contrasts are greater in spring and fall than in mid-summer, and the frost-free period may be shortened in many years by surges of cold, polar continental air masses. The average frost-free period is over 120 days in the lowlands of southeastern Alberta and in a smaller area in southern Manitoba. The northern and western fringes of widespread agricultural activity are in areas with 60 to 70 frost-free days. The growing season is considerably longer. In southern Alberta it extends from mid-April to late October and, in the northern Peace River valley, from early May to late September.

One of the best indices of the heat supply available for plant growth is the number of degree-days above 42°F (5.6°C) (Figure 1.5). The range is from over 3000 F° days (1667 C° days) in the Medicine Hat area of Alberta and the Morden area of Manitoba to less than 1000 F° days (556 C° days) near Hudson Bay. Crops such as sugar beet and grain corn may be grown in the warmest areas; small cereal grains are not widely produced with less than 1800 F° days (1000 C° days) and only a limited number of vegetables and forage crops are produced in the cooler areas. Forest growth is roughly proportional to heat receipts; thus a conifer stand for pulp production can be produced in 80 years with 2250 F° days per year or 120 years with 1500 F° days per year. Some allowance should be made for day-length variations with latitude because this is an important positive factor in northern areas where heat supplies are marginal. Conversely, growth is inhibited by severe frost within the growing season, by moisture deficiencies, and by the occurrence of hail or strong winds, so the degree-day index must be used with care.

The average annual precipitation patterns of the prairie provinces reflect the average annual levels of orographic, cyclonic, and convectional storm activity (Figure 1.6). The heaviest precipitation is received in the higher mountain areas of the southwest, where winter snowfall greatly exceeds summer rains. Most areas to the east of the mountain zone have a more continental regime with about 70 per cent of the total precipitation falling in the April-September half of the year. The greater part of the plains precipitation falls from Pacific air masses in relatively frequent showers, which are usually associated with cold frontal or summer convectional uplift. The Pacific air masses lose much of their moisture over the western cordillera but they release additional moisture if cooled suffi-

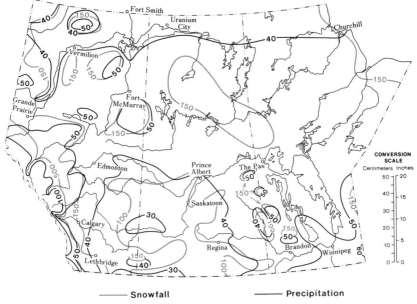

——— Snowfall   ——— Precipitation

1.6
Mean Annual Snowfall - 1931-1960-Cm.
Mean Annual Precipitation - 1931-1960-Cm.
(Adapted from Longley)

ciently in strong uplift situations. Low-pressure centres pass from west to east. In winter, their tracks are usually in the United States, but they tend to shift to the southern half of the prairie provinces in spring and fall, and to the northern half in summer. The tracks fluctuate appreciably, however, and the local precipitation patterns fluctuate with them.

The presence of cold air aloft or of heavy surface heating often results in convectional showers if the air masses are relatively humid. Orographic intensification, especially of cold-front showers, occurs in upland areas within the plains as well as in the foothills and front ranges of the Rocky Mountains. In summer, warm, moist tropical marine air moves northward from the Gulf of Mexico and relatively heavy rains are experienced, especially in the southeast. The depth and frequency of penetration of these air masses vary greatly. In some years the widespread 'million dollar' rains provide moisture for excellent crop growth, and in other years there is little movement of these air masses into the region. These air masses rarely reach the prairies at surface level in winter, but warm-front cloud cover aloft is more often present in the southeast – with

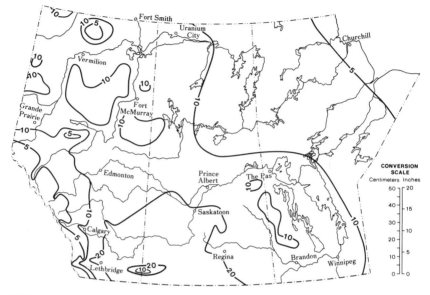

1.7
Average Water Deficiency (Thornthwaite, 10 cm. Storage)
(After Laycock 1967)

snowfall. The opinions of meteorologists differ greatly, but the proportion of annual precipitation from mT air may be roughly 50 per cent for
Winnipeg, 25 per cent for Medicine Hat, and 10 per cent for Edmonton.

Winter snowfall is relatively light in the southwestern plains and, with
the frequent chinooks, there are usually widespread bare soil and grassland areas there and in the foothills during the winter. Winter grazing and
browsing by cattle and game animals is therefore possible, except in some
years of heavy snowfall and limited chinook activity, when mortality
rates are high if winter feed supplies have not been provided. This is the
critical period for many birds and animals. In all areas, the melting of
winter snows is important in soil moisture recharge and runoff patterns
and both vary greatly because of variations in snowfall amount, drift
distribution, winter melting and evaporation, spring weather infiltration
capacities, frost patterns, and surface detention storage patterns – some
of which can be modified by tillage practices.

The precipitation of most of the agricultural areas of the plains is
exceeded by potential evapotranspiration, and moisture deficits are usually
experienced from mid-summer to early fall (Figure 1.7). The deficiency
patterns are closely reflected in vegetation, soil, and agricultural patterns.

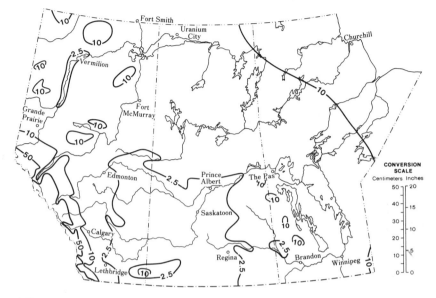

1.8

Average Water Surplus (Thornthwaite, 10 cm. Storage)

(After Laycock 1967)

Summer-fallow in alternate years is employed to reduce deficiencies by several inches in crop years. Thus, cereal crops can be grown well into the drier plains – for example, where average deficiencies exceed 20 centimetres – although some risk is involved. If much wetter than normal conditions are experienced, large yields may be produced; if dry conditions prevail, yields may be very low even after fallow. A small deficit in late summer during the crop ripening and harvesting period may be helpful. Locally, if moisture storage is below average, deficits and surpluses are both greater. Similarly, in forest areas the greater depth of soil moisture utilization results in smaller deficits and surpluses than those indicated in Figures 1.7 and 1.8. Deficits in forest areas, particularly in sequences of dry years, can be related to patterns of fire frequency, intensity, and extent.

Many other aspects of climate affect natural and cultural patterns. Most of the more settled areas have 2000 to 2300 hours of bright sunshine per year, which are high totals for the latitude. The drier southwestern plains have the highest totals, but they also experience slightly higher winds than most other areas. Hail damage is great in some years, particularly in the western plains. Fall rains during the harvest season are hazards in the northern and eastern agricultural areas, but some delayed spring harvests have been moderately successful. 'Cyclones' and tornadoes, such as that in Regina in 1912, can cause local damage but they are very infrequent. In marginal areas particularly, microclimatic variations can be very significant. Thus minor air drainage variations may cause a low marshy area to have a frost-free season that is a month shorter than adjoining areas relatively few metres higher.

Water Resources

The water surplus patterns reflect precipitation and evapotranspiration patterns closely, and with allowances for seasonal concentrations and storage, local water supply is a variable residual of these two major variables (Figure 1.8). In the drier plains the average surplus is less than 2 cm per year, yet it may range from zero in many years to as much as 12 cm in the wetter years. Local snow drifting and relatively impermeable surfaces (bare rock, roads, or frozen clay soils) often result in above average local yields and flashy snowmelt or rain runoff. Sand plains have good groundwater recharge and a more dependable flow from springs. Clay plains have poor groundwater supplies because the larger supplies from heavy rains and above normal snowmelt run off in surface flow, and the balance is stored within plant root depth and is lost to transpiration in summer. Many streams have low yield and very flashy flow for little more than the snowmelt period and following unusually heavy rains. Flow variation from year to year is accentuated by variations in the area of the contributing basin, many of the local depressions overflowing only in the wetter years. Most of the local water bodies serve as evaporation basins for surface and groundwater inflow and contribute little or no groundwater flow to streams that pass out of the region. Some of the larger lakes do not have external drainage and many others do not have outlets in the drier years.

Many of the patterns noted are present in the more humid plains, but in general these regions have better groundwater recharge and a better base flow is available for more of the year in the larger streams. Seasonal regime and year-to-year changes are still very pronounced. More of the depressions contain water throughout the year and drainage is more of a problem in spring in agricultural areas. Conflicts of interest between those who would protect wildlife, including the very large numbers of migratory waterfowl, and those who would drain land for more intensive agricultural use are growing. Unfortunately, the generalized claims are usually too sweeping, and many compromises involving allocation of water bodies that are best suited to each purpose, controlled drainage, and supply consolidation could be mutually beneficial.

In dry years about 90 per cent of the streamflow of the North and South Saskatchewan basins originates in the Rocky Mountains and foot-hills, which have only 15 per cent of the total basin area; in wet years the proportion is about 70 per cent. The importance of the headwater region is due also to the greater dependability, better seasonal distribution, and better quality of flow. Watershed management is important, primarily for regime and quality improvement and for erosion and flood limitation, but also for local recreational, wildlife, forestry, commercial grazing, mining, and other uses, and many use conflicts must be resolved. The high yields can be attributed to high precipitation, low potential evapotranspiration, and widespread bare rock. Yield increases from present levels are unlikely because most of the forest is young following major fires and cutting, and it will consume more water as it matures. Artificial storage for regime improvement for hydropower, urban, industrial, irrigation (late summer), and other demands is in partial conflict with recreational and wildlife use of the region.

Because of its bare rock and shallow soil surfaces, the Shield has larger surpluses than indicated in Figure 1.8, which is based on a 10 cm average soil moisture storage. Runoff into the lakes and marshes is large, and the natural detention storage contributes to a relatively flat flow regime for most rivers. Hydropower development is favoured also by the natural head concentrations, the excellent footings for dams and spillways, the lack of flooding, and the high water qualities.

The major streamflow rises in northern, eastern, and mountain areas, but the water demands are concentrated in the plains. The Slave River with close to 100,000,000 acre-feet of annual flow (4000 m³/sec), the Nelson with 65,000,000 (2500 m³/sec), and the Churchill with 30,000,-000 (1200 m³/sec) are much larger than the North and South Saskatche-wan, each with 7,000,000 (270 m³/sec) and the Red with 5,000,000 (200 m³/sec). Winnipeg is well supplied from several directions, but the plains to the west depend largely upon mountain sources. With growth in demand there must be increasing review of the alternatives in water management – those involving supply augmentation including watershed management, desalination, weather modification, and inter-basin transfer from north-flowing streams and those involving re-use, more efficient use, reallocation of supply, non-expansion in demand, industry movement to supply, and other possibilities. No one solution should be stressed to the exclusion of others, and the hazards, consequences, and opportunities

Grassland

Parkland

Mixed Forest

Coniferous Forest

Alpine

Tundra Transition

Tundra

0		100		200 Miles
0	100	200	300 Kilometers	

1.9
Vegetation
(Adapted from Atlas of the Prairie Provinces, Atlases of Alberta, Saskatchewan and Manitoba.)

of each action must be considered. Of the physical alternatives, transfer has the brightest potential, at least partly because a part of the transferred water might be sold as an export commodity to the United States. This large renewable resource of northern areas might well be their most important one in the future.

Vegetation

The vegetation patterns of the prairie provinces are zonal, corresponding in general to climatic and soil patterns. There are many local variations in relation to physiography, water supply, and such disturbances as climatic fluctuations, fire, cutting, cropping, and grazing. The greatest changes have occurred with cultivation in the grassland and parkland areas and with fire in the mixed forest areas (Figure 1.9).

Grasslands occupied much of the warmer and drier southern portions of the plains, and have provided the region with its popular name. The term 'prairie provinces' gives a false impression, however, because grassland was dominant in only one-sixth of their area and only a small part of that was an extension of the 'prairie' of the United States. Within the region, the tall-grass prairie west of the Red River is differentiated from the short-grass prairie of the drier plains and the mixed-grass prairie which is transitional to the parkland. This differentiation is only partly on the basis of species composition. The tall grasses of the first zone are characteristic of poorly drained sites on the Lake Agassiz plain, but mid-grasses, such as spear grass, are dominant where drainage is better. Fringes of woodland species more common in the adjoining parkland are also present along the Red River. A regional moisture deficit of about 15 centimetres is largely overcome by surface supplies and soil moisture storage.

The short-grass prairie is dominated by blue grama, June grass, wheat grass, and spear grasses which have a stunted growth in most years. The average moisture deficit is 20 to 30 centimetres. There is quick response to extra moisture and the mid-grasses appear to dominate in the wetter years. Similarly, the wetter sites with concentrations of groundwater and surface runoff and with extra moisture from drifted snow have more luxuriant growth. Tree and shrub growth is largely phreatophytic or is located in upland areas where it is favoured by additional precipitation and smaller evapotranspiration requirements, particularly on north-facing slopes. The drier sites have relatively sparse growth and even xerophytes, such as prickly pear cactus. There are many complex associations relating to particular drainage, soil, topographic, and other variations. In saline areas, where groundwater that has passed through marine deposits comes to the surface and evaporates leaving 'alkali' precipitates, the more salt-tolerant species are present in association with a wide range of plants that thrive on the fresher groundwater, or are tolerant of seasonal flooding. In sandhill areas, the higher stabilized dunes are relatively dry in summer because of the low retention storage capacities of the soils. The spring surpluses percolate to lower levels and water tables may be high, creating sloughs in depressions. The cover range relating to these extremes of moisture supply is frequently very great. In many of the rougher moraine and till plain areas the 'pothole' sloughs in swales and kettles frequently serve as evaporation basins for groundwater supplies. They support a diverse marsh cover with rings of trees and shrubs on their margins where groundwater flow can be intercepted by the deeper roots. On flatter terrain, the water supplies are more variable and summer evaporation is rapid: awned sedge and reed grasses are charactertistic species in the shallow depressions. The river valley complexes are among the most diverse, including many trees and shrubs that are more common in adjoining wooded regions.

The mixed-grass prairie has a denser, taller, and more diverse cover, but it is transitional from both long and short grass prairies to parkland rather than having a mixture of tall and short grass species. Both are present in different areas but mid-grass species predominate. The moister areas have scattered groves of trees, largely aspen, separating mixed grasses which grow more luxuriantly and with a smaller proportion of forbs than in drier areas. Strips of trees and shrubs spread well into the grassland region along the valleys of present and pro-glacial streams, and clumps are widespread wherever there is some concentration in moisture supply. Winter snow-drifting patterns add to the diversity, the open areas losing a part of their annual moisture supply to clumps and groves of trees in which snow drifts are deepest. Man has contributed to some of these differences by clearing fields for cultivation and leaving trees and shrubs along fence lines and road allowances. Vertical zonation is also present; thus the higher-level grasslands of the Cypress Hills include dense fescue growth and many more trees and shrubs than the adjoining lowlands. Aspect is also important; north-facing slopes have shrub and tree cover, including some conifers, while south-facing slopes have a sparser grass cover.

The parkland areas have a mixture of grassland and woodland cover. Grassland is more prevalent on the warmer and drier margins, and trees on the cooler and moister margins. Average moisture deficits of 10 to 20 centimetres are experienced, but the frequency of intense prolonged drought and other factors appears to have been at least as important in determining natural vegetation boundaries. Trees may be destroyed in successions of dry years, and recovery is slower than for grasses. Fire, grazing, and browsing and the tendencies of trees and shrubs to improve their sites for future growth, by concentrating snowmelt supplies, have been important. A continuous tree cover might not be supported but a mixed cover may be, because of local differences in snow cover, soil moisture retention storage capacity, runoff, groundwater supply, and evapotranspiration relating to aspect. Aspen poplar predominate in most parkland groves but bur oak and other Great Lakes Forest species are present in Manitoba parklands and various montane and subalpine species occur in the Alberta foothills. Many of the wetter sites have mixed forest species such as balsam poplar and an occasional spruce. Many shrubs, including saskatoon, dogwood, and willow, are present under different moisture conditions. The fescue grasslands are among the richest present, but intensive grazing has permitted less nutritious and less palatable species to become more common.

The mixed forest regions usually contain both conifers and deciduous broadleaf species, but repeated fires have contributed to the development of many even-aged stands of aspen and some stands of balsam poplar. The average moisture deficit is 5 to 15 centimetres, but severe droughts are frequent and fires then spread rapidly. Here and in the coniferous forest areas there is a patchwork mosaic of age classes and sometimes species association relating to past fires. A succession to conifers is likely if fires are limited in the future but growth is slow (80 to 120 years for pulp-wood). Patches of mature conifers include lodgepole pine in foothill areas, jack pine, white and black spruce, tamarack in wet sites, and smaller areas of white cedar and red pine in eastern Manitoba. Shrubs such as hazel and highbush cranberry and patches of wet sedge meadow are wide-spread. The wetter 'muskegs' or bogs contain peat mosses and low shrubs such as Labrador tea. Large areas are poorly drained, at least partly because water surpluses are not large enough to erode a dense, incised drainage net. They remain on relatively flat surfaces long into the summer and wet organic mats are very extensive, particularly on lacustrine plains.

The coniferous forests are widespread in the more humid regions with enough heat for tree growth. The boreal forest changes slowly in species composition from east to west. In much of Manitoba and Saskatchewan it is on Shield surfaces where there is little uniformity in soil, slope, or related cover. Summer moisture deficits of 5 to 12 centimetres are exceeded by spring snowmelt surpluses but droughts can be intense, par-ticularly on the drier margins. Marshes and lakes are numerous and the relatively bare rock supports little growth. In drift areas the growth is better but many parts of these have mixed forest cover. Widespread fires have contributed to the growth of extensive stands of jack pine but spruce are dominant in most areas. White spruce and balsam fir are widely present on well-drained sites and black spruce and tamarack share the wetter sites. White birch and aspen poplar often follow fires and are succeeded in time by conifers on most sites. Sphagnum peat bogs and muskeg mantle many of the depressions between rock outcrops. Some of the coniferous forests to the south and west of the Shield are on sandy lands where jack pine dominate the fire succession growth, or on extensive deltas where some of the better-drained levees have excellent white spruce in narrow, protected strips plus black spruce and tamarack extending out into the marshes and in cool or wet sites where fires have not recently contributed to the growth of deciduous stands.

The coniferous forests in and near the Rocky Mountains include lodgepole pine rather than jack pine, and Engelmann spruce and alpine fir rather than balsam fir. In the southern mountains there are no black spruce and few white birch but Douglas fir, limber pine, and alpine larch are present on different sites. Frequent fires in the drier valleys and foothills have contributed (with cutting) to the widespread presence of lodgepole pine and various poplar. Some of the drier valleys and southern foothills have productive fescue grasslands, and the alpine complex includes high-level alpine meadows above the tree line (2000 to 2500 metres depending partly upon exposure and soil availability).

The northern transition to the tundra has a more open parklike and somewhat stunted forest growth. The spruce are separated by low shrubs and mats of lichens. White spruce are present only on the better sites and black spruce predominate. Jack pine and white birch frequent the recently burned sites. Drainage and deep rooting are inhibited by permafrost and the extensive marshes of the Hudson Bay lowland have only sparse and stunted tree growth. The upper portion of the Caribou Mountains in Alberta is largely sphagnum moss muskeg with a limited growth of black spruce.

A narrow fringe of arctic tundra is present near Hudson Bay. Stunted birch and arctic willow are present in poorly drained areas of lichen, mosses, grasses, and sedges. Stunted trees extend as far as the coast along streams where some additional heat and better drainage conditions are found.

Animal Life

The wildlife ecosystems are as varied as the vegetation patterns, and a more diverse range is supported than in any other part of Canada. In the grasslands and parkland areas the bison were very numerous but numbers declined very rapidly with heavy hunting in the period 1870–85 and the relatively few that now remain are largely in parks. Pronghorn antelope, mule deer, and white-tailed deer are more numerous. Northern forests have moose in the wetter areas, caribou, and growing numbers of deer. Alpine areas have wapiti, mountain goat, and bighorn sheep. Large predators such as mountain lion, grizzly and black bear, and wolf may be found in different regions. Smaller animals including many fur-bearers such as beaver, muskrat, mink, fox, and squirrel are numerous in northern forest and marsh areas. Bird life is also diverse and seasonal migrations add to the variety. Prairie potholes and fields support many millions of waterfowl that fly southward each fall and return in the spring. By this means, much larger numbers may be supported than for most other species which remain through the winter. The lakes and streams contain significant numbers of commercial and game fish. The lake area is large but growth is relatively slow and the potential for production is only moderate. Hudson Bay has relatively few fish and only the white whale or beluga is caught in sufficient numbers for commercial production in some years.

Soils
The soil patterns of the prairie provinces reflect to varying degrees the influences of the soil-forming factors – parent materials, slope and drainage, climate and vegetation – acting through time and as locally modified by human use (Figure 1.10). The major parent material patterns reflect particularly the effects of glaciation and post-glacial erosion and deposition, but surficial materials vary also with bed-rock and organic buildup patterns. The time period since glaciation has been short and most soils are intrazonal or azonal, so that parent material, slope, and drainage differences are reflected to a greater degree than is generally true for unglaciated regions with less climatic change in recent time. The effects of climate and vegetation are becoming more and more pronounced with time and, with numerous local variations, a general zonation has been established. With climatic changes toward moister and possibly cooler conditions in recent centuries there has been some advance of various types of tree cover into the grassland, and some degradation of grassland soils on the forest margins has occurred.

The Chernozemic Order includes the Brown, Dark Brown, Black and Dark Gray great groups of soils. These have developed under grass cover (with associated forbs and shrubs) in semi-arid and dry, sub-humid climates. In virgin soils the Ah horizon should be thick (not less than 9 cm) and dark. It has 1.5 per cent to 30 per cent organic matter and well-

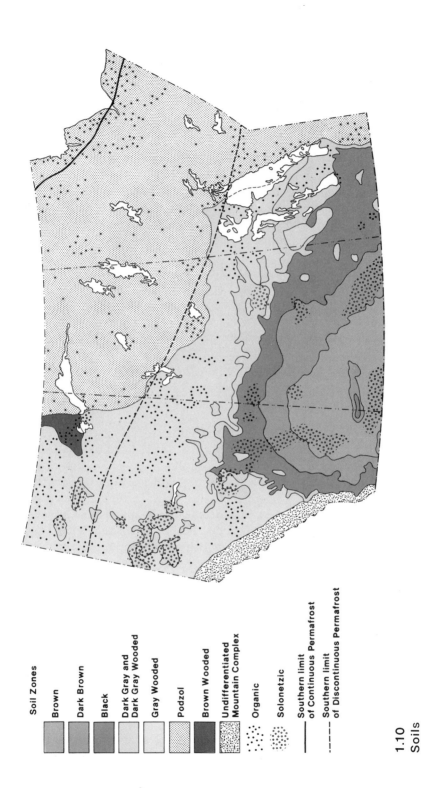

Soil Zones

Brown

Dark Brown

Black

Dark Gray and
Dark Gray Wooded

Gray Wooded

Podzol

Brown Wooded

Undifferentiated
Mountain Complex

Organic

Solonetzic

——— Southern limit
of Continuous Permafrost

– – – Southern limit
of Discontinuous Permafrost

1.10
Soils

(Adapted from Atlas of the Prairie Provinces, Atlases of Alberta, Saskatchewan and Manitoba.)

flocculated, moderate to strong structures that do not become massive on wetting or single-grained on drying.

The Brown or Light Chestnut soils are associated with the drier grasslands. They are relatively shallow, light-toned, and low in humus for Chernozemic soils, and a Ck horizon of lime accumulation is usually present at 35 to 50 centimetres. Approximately 25 per cent of the soils in this zone are cultivated, largely in lacustrine and bevelled till plains where the moisture retention capacities of the clay soils and the topography are most favourable. Native grassland pasture is more widespread. Erosion by wind and water are hazards if the organic content of surface soils is depleted greatly with use. Summer-fallowing, stubble mulch tillage, restricted grazing, and other conservation practices are becoming customary. These soils are very productive with irrigation and with the heavier rains of the wetter years.

The Dark Brown or Dark Chestnut soils are usually associated with and developed from the mesophytic grasses and forbs. They are darker and deeper, and have more organic matter and deeper Ck horizons (about 60 cm) than Brown soils. About half the soils of this zone are cultivated, again particularly in the better soil moisture retention storage areas with topography that is well adapted to large-scale, highly mechanized farm operations.

The Black soils are associated with a mixed grassland-parkland cover of mesophytic grasses and forbs with discontinuous or thin tree and shrub cover. They have Ah horizons of 12 to 25 centimetres in average depth ranging from the drier 'thin black' soil areas to the more humid sites. The average organic matter content is about 8 per cent and the Ck horizon is relatively deep (about 75 to 120 cm) and less pronounced in its development. About two-thirds of these soils are cultivated and there is less focus upon the finer-textured, flatter terrain areas than in the previous zones although they are still favoured. Soil moisture conservation practices are less necessary but with the more humid conditions and intensive use, there is good response to fertilizer additions.

The Dark Gray and Dark Gray Wooded soils are transitional from the Chernozemic to the Luvisolic Order. They approach or slightly exceed the requirements of a Chernozemic Ap horizon (cultivated 15 cm) in depth, colour, and organic content. Both have originally developed under grass, but they are now in the more wooded parkland and mixed forest transition areas. If the Ae (eluviated) horizon is more than 6 centimetres in thickness and a Bt (clay accumulation) horizon is present, the soil is considered to have undergone sufficient podzolization to be classified Dark Gray Wooded and in the Luvisolic Order. There is usually a thin leaf mat upon the surface and the lime concentration, if present, will be 100 to 150 centimetres below the surface. In Manitoba a significant part of this zone has 'rendzina' soils that are high in lime and meet the soil requirements despite the more humid conditions. About one-third of the area with these soils is cultivated, and drainage and frost limitations are usually greater than those relating to moisture deficiency.

The Luvisolic Order includes the Gray Wooded soils, which are more widespread than those of any other great group in the prairie provinces. It consists of soils that have developed under boreal and mixed forest, generally on basic parent materials. The Orthic Gray Wooded soils have organic L-H (surface) horizons, light-colored Ae and Bt horizons. The Ah horizon, if present, is less than 5 centimetres in thickness. The cultivated Ap horizon is usually moderately light-toned. Less than 10 per cent of this area is cultivated, largely because of frost and drainage limitations, but these soils can respond well with certain forage and grain crops in most years.

The Podzolic Order includes several Podzol soils which are present in the more-humid, largely Shield areas of the north and east. They have developed under coniferous and mixed forest and heath cover, largely on acidic parent materials. They have organic L-H horizons (largely needle leaf litter), some have thin Ah horizons, and most have leached (Ae) horizons over B horizons that contain accumulations of organic matter and iron and aluminium. The soils are acidic and few are cultivated. The widespread bare rock and poor drainage result in less than zonal development in most areas.

The Brunisolic Order is represented by Brown Wooded soils which have developed under mixed forest, grass, heath, and tundra cover and have a brownish Bm horizon. They are virtually uncultivated but some provide pasture for bison.

The Solonetzic Order and Solonetzic and Solodic intergrades of Chernozemic and some Luvisolic soils are widely distributed. Most are in grassland areas or where grasslands have been present, and they have developed where the parent material was more or less uniformly salinized. Marine bedrock tills and groundwater salt precipitate areas are most likely to have them. The hard columnar structure of the saline Bnt horizon is a distinguishing feature. The A horizons are usually shallower and more subject to wind erosion than those of the adjacent orthic soils. Cultivation is accordingly less widespread but many are cultivated.

The Organic Order consists of soils that have developed dominantly from organic deposits. Most are saturated for much of the year and contain 30 per cent or more organic matter in surface horizons. They are widespread in the more humid regions and few have been drained or cultivated, largely because better soils are still available elsewhere.

Gleysolic and Regosolic soils are widely distributed. They are associated with impeded drainage, with bedrock exposures, and with areas where erosion and deposition have been too recent for significant profile development.

The 'Undifferentiated Mountain Complex' soils contain virtually all of the categories listed previously, from the Brown soils of dry mountain valley grasslands to Podzols in the wetter uplands. The azonal and intrazonal variations relating to high relief and diversity in soil-forming factors are most widespread.

Permafrost in northern areas has been an important factor in the limitation of drainage and the development of organic soils and the lack of development of others. On its southern margins and in alpine areas it is associated largely with peatlands and north-facing slopes in uplands. It varies in thickness from relatively few centimetres to 60 metres in the discontinuous zone. The active surface layer becomes shallower with organic buildup and shade and tree growth is inhibited.

Evaluation

The current use of the physical resources of the prairie provinces is largely indicated by the net value of production (Table 1.1). In the past the foci of production have been upon single-use industries oriented to export in a largely unprocessed form: furs, wheat, oil and gas, potash and nickel are leading examples. This is still largely the pattern, with more and more products entering the export list, but there is also a growing level of production for local use and of processing for both local and external markets. Integration is still not advanced but the basis for it is improving.

The prairie provinces account for 50 to 60 per cent of the net value of Canadian agricultural production, even in a climatically poor year such as 1968. The longer-range trend is toward a growing proportion, largely because the region has over three-quarters of the occupied agricultural land of Canada and this proportion is increasing. The landforms in terms of slope, drainage, and parent material are very favourable and the potential for expansion and more intensive use is better than anywhere else in Canada. In eastern Canada, for example, the abandonment of submarginal land is proceeding rapidly. The climatic resources of the plains, in terms of length of growing season and heat and moisture supplies,

Table 1.1 Census value added for goods producing industries – by province and industry, 1968 (in thousands of dollars)

	Manitoba	Saskatchewan	Alberta	Prairie provinces as percentage of Canadian total	Western Canada as percentage of Canadian total
Agriculture	271,748	644,581	581,468	52	57
Forestry	2,111	5,097	6,348	2	53
Fisheries	3,276	1,382	917	3	34
Trapping	1,601	1,551	1,730	40	54
Mining	113,526	300,136	1,019,726	45	56
Electric power	62,717	59,436	85,133	15	27
Total primary production	454,979	1,012,183	1,695,322	38	51
Manufacturing	442,994	169,928	606,032	7	15
Construction	280,234	290,678	665,220	23	37
Total production	1,178,207	1,472,789	2,966,574		

Source: *Canada Yearbook*, 1970–1.

are also good. Small areas in British Columbia and Southern Ontario are better for some crops but the much larger areas with favourable (if variable) conditions in the prairie provinces give them a distinct natural advantage for greater long-term production. No other part of Canada is as well endowed with natural grazing lands, and the soil resources are by far the best in quality and extent. Farm consolidation is sometimes painful, and few small-scale operations are needed, but cultivated acreage and unit-area production will both continue to increase greatly in the future.

Forestry is relatively minor. The timber reserves are only about one-eighth of those of Canada and one-quarter of those of British Columbia. Moreover, much of the timber is small and slow-growing and the natural fire history makes long-term development precarious. Because of its extent, however, it is still a very significant resource, and current pulp-mill development ensures that its exploitation will increase.

The commercial fishing and trapping industries are small and have little potential for growth. The value of the fish and wildlife resource, however, is only partly indicated by these data. Recreational fishing and hunting are still expanding rapidly, and the concern for wildlife conservation is a growing force. With greater stress on wildlife in multiple-use planning of land and water, this resource should be more effectively utilized.

The mining industry has passed agriculture in net value of production in recent years, largely because of the great growth in oil and gas. In 1971 the prairie provinces contributed over half of the net value of mineral production in Canada. Continued growth is expected, partly because of oil sands development and greatly expanded activity in coal mining. The Shield production of metals is growing rapidly and exploration is most promising. Potash and materials for an active construction industry are growing in output. There is good reason to anticipate continued growth although these minerals are non-renewable resources and supply depletion will result in longer-term declines for some. Conservation, the requirements of increased processing and associated production in the region, and a growing stress upon renewable resource use will be important features in future development.

Electric power production is growing, largely in response to market demand. Hydroelectric power development is continuing and the potential is still significant in northern areas, but thermal power production is now much more important. Coal, gas, and diesel plants have relatively low-cost sources of supply in the south and west. The advantages once held by eastern Canada and British Columbia, and by Winnipeg in the prairie provinces, with their low-cost hydropower have been lost as demand has exceeded supply and the advantage now rests with areas having cheap fuels – unless fuel movements and nuclear power development are heavily subsidized.

Water resources are large, particularly in northern, eastern, and mountain areas, yet less than 6 per cent of Canada's streamflow rises in the prairie provinces. Additional supplies rising in British Columbia, northwestern Ontario, and the United States pass through prairie areas. Western Canada has well over half of the national flow, and this is also a greater flow than that rising in the 48 conterminous states of the United States to the south. Topographic patterns are favourable for diversion. The hundreds of thousands of lakes, ponds, marshes, and muskegs provide varied opportunity for recreation and habitat for wildlife – uses that are of rapidly growing importance.

In summary, the region is very well supplied with natural resources and has a disproportionate share of Canada's total. The surprising point is not that it produces so much: it is rather that, with its resource advantages, it does not contribute an even larger share of the national production.

The diversity in the physical landscape is increasingly evident in the variety of uses which are made of it. It is not just forest, rivers, and lakes as it must have seemed to the early fur traders, or flat and open grassland as much of it must have appeared to pioneer farmers intent upon producing wheat. It is becoming increasingly diverse in its development and the old bonds of common commodity production for export are breaking down. It is not a single 'prairie' unit – and never was. It is a changing complex to residents and visitors alike.

2 Some Reflections on Man's Impact on the Landscape of the Canadian Prairies and Nearby Areas

J.G. NELSON

This essay aims to provide an understanding of the relative recency, the depth, the scale, and the acceleration of human impact on the landscapes and ecosystems of the Canadian prairies and nearby areas. The words landscape and ecosystem are used interchangeably and comprehend such interconnected elements and processes as bed-rock, soils, vegetation, wildlife, climatic change, fires, animal cycles, and man. The terms are impossibly comprehensive but reflect the complex nature of the management problems resulting from human interference with nature.

Origins
The interests of the landscape historian must verge on the timeless. Some of the rocks of the prairie provinces are Precambrian and billions of years old. Ancestral forms of the conifers of the prairie fringes and highlands originated in the Palaeozoic, hundreds of millions of years ago. Beaver and other fauna date from the late Tertiary, several million years ago. About one and one-half million years before the present, glaciers began their complex play of advance and retreat which was to shape so much of the modern landscape. The ice also caused fundamental changes in animal life. The formation of the glaciers led to the withdrawal of water from the oceans and the emergence of land bridges, like that of the Bering Straits across which deer, elk, bear, mammoth, and other fauna migrated to North America. Man also used this route to make his first entry into the New World. About ten thousand years ago many of these animals became extinct, possibly because of warming and other climatic and vegetative changes at the end of the Pleistocene, and possibly because of hunting by man.

With the retreat of the ice and the gradual uneven warming of the climate over the last 12,000 years, vegetation has undergone a variety of changes in distribution and character. Pollen analysis suggests that the first vegetation to grow in parts of the central and eastern prairies included large quantities of spruce, and, in southwestern Alberta, lodgepole pine.

As the climate warmed, notably during the Altithermal of 8000 to 4000 B.P., conditions worked against the maintenance of forest and in favour of grassland and parkland. The major changes probably emanated from a combination of increased aridity and fire. Lightning and man-made fires have been normal in the plains for the last two hundred years and there is no reason to doubt their presence for the last several thousand years.

The natives of the plains used fire for a variety of purposes, including signalling, hunting, hiding a trail from enemies, and warfare. Fires also were caused by accidents. These arose, at least in part, because of the technology available to produce and care for fire. The fur trader David Thompson observed that when the Piegan shifted camp, 'a careful old man or two' was entrusted with the fire, which was carried to the next camp in 'a rough wooden bowl with earth in it.' The historical literature on the northern plains also contains frequent mention of fires left unextinguished by the natives. The initial tendency is to view this as carelessness but many of the fires could have been banked and left for later travellers. How frequently these fires would escape into the grass is impossible to say, but such accidents are mentioned by fur traders.

In the days of the unfenced open grassland, fires could burn for weeks and cover thousands of square miles. They would also have different effects under different physical circumstances. In some cases the balance between fire, aridity, and other influences would result in the elimination of all trees and the creation and maintenance of grassland. In other areas, the hardy aspen poplar might be able to regenerate from its subterranean root system quickly enough, between fires, to maintain the scattered groves typical of the savannas of the northern grassland fringe.

In pre-European days, vegetation was also influenced by the activities of numerous and varied animals. Estimates of the bison population range between twenty and forty million for the Great Plains as a whole. About a quarter of the animals have been envisioned as living in the northern plains of the United States and Canada, along with millions of antelope, elk, deer, ground squirrels, foxes, plovers, grouse, ducks, cranes, and other fauna. Many of these animals had considerable effect on the landscape by feeding, walking, and wallowing. For example, by its repeated tramps to water, the bison left many deep trails down valley slopes, some of which became a focus of runoff. The bison also frequently wallowed or rolled in the mud or in dry silt, clay, or sand, especially in the spring, when the loss of its pelage exposed it to mosquitoes and other insects. The resulting wallows were commonly 5 or 6 metres wide and 12 metres long, and often coalesced to form compound, overlapping basins of exposed soil that stretched for kilometres. On present-day bison or wildlife reserves, the wallows promote sheet wash, water erosion, and possibly some wind erosion, although there are few reliable observations of the last.

The bison also grazed heavily. Numerous references can be found to bison moving over the land in thousands or tens of thousands, eating the grass short as they went along. John Palliser, who explored the Canadian plains in the late 1850s, complained that grazing was sometimes so heavy that feed could not be found for his horses. The effect, presumably, would have been to reduce the sensitive mid to long native grasses, such as rough fescue and spear grass, and to favour the relatively resistant mid and short grasses, such as wheat grass, grama, and June grass. In other words, short grass was probably more widespread than at the beginning of permanent agricultural settlement, when the animal populations had been sorely reduced by several decades of intensive hunting. It follows that the picture of the parkland and much of the grassland as a relatively undisturbed sea of mid to long grasses, stretching to the horizon, has been overdrawn. Such a landscape probably could not have developed until the grazing pressure was lessened, thus allowing fescue and other grasses to become more frequent and to grow up to the bellies of the horses of enthusiastic early settlers. Undoubtedly some snowberry or other shrubs or forbs also began to cover the exposed wallows and bison trails. It was this changing landscape that the early settlers saw, and the one that is perceived to be truly reflective of 'pristine' nature as it was prior to the 'overgrazing' and other 'destructive' processes of the white man.

This does not mean that the long fescue grasslands of areas such as the Porcupine Hills, south of Calgary, may not have existed before the arrival of the European. Indeed, the suggestion has been made that such grasslands could be maintained if the grazing of bison and other animals were concentrated in the winter when fescue is less easily damaged or destroyed than in summer. In the fur trade and exploration literature of the eighteenth and nineteenth centuries, reference is made to large numbers of bison in this foothills area in winter. Today the warm, dry chinook wind is especially frequent and effective there and, if the same were true in the past, it would have tended to keep the snow shallow, thereby providing good feeding conditions. In spring and summer the bison might have moved onto the plains, so reducing the pressure on the fescue at its most sensitive season. Unfortunately, few traders and explorers passed through the Porcupines country in summer, so that it is difficult to know how numerous the animals were then.

The bison has also been envisaged as preventing the growth and extension of woody plants onto the grasslands. However, the animal generally does not seem to eat much young aspen and willow today, so that its effects through feeding were probably less than through trampling and rubbing. Browsing animals such as the elk probably had a greater effect in preventing the extension of poplar. Elk were exceedingly numerous over much of the North American plains in early European days. Peter Fidler and Lewis and Clark frequently noted them along the South Saskatchewan and the Missouri in the late 1790s and the early 1800s, when they appeared to be present in almost every poplar grove. Elk and deer both browse on snowberry, buffalo berry, and other shrubs, as well as on the succulent buds and tips of the willow and poplar. Such browsing undoubtedly killed or damaged large numbers of trees and limited the growth and expansion of the groves. These effects could have been reinforced by the moose, whose range then extended much farther south. It was a favourite food of the trader, and was shot in the hundreds in the 1750s and 1770s in central Saskatchewan and Alberta. The number of the moose undoubtedly was also affected by changes in the extent of suitable habitat. The fur trade records suggest that moose tended to be numerous where beaver were found.

The beaver, of course, was a major modifier of landscape, a rival to the bison, although its effects were concentrated in the stream valleys. It built dams and created extensive ponds which flooded large areas near the Missouri and its tributaries, as well as the Red Deer, the Battle, the Carrot, and other streams. Such ponds promoted the growth of vegetation upon which the moose like to feed, particularly in summer. When the beaver were hunted down or eliminated, and their dams or ponds destroyed or deteriorated, there was undoubtedly an effect on moose habitat and numbers.

In other ways, faunal actions promoted the spread of trees. The aspen today is generally spread by suckering roots and not by seed, which has difficulty penetrating the grass sod. However, the seed will establish itself where fresh mineral soil is exposed by burrowing animals and where some protection from disturbance is afforded by other plants. In earlier days, the wallowing of the bison, and digging by ground squirrels and other animals, exposed large areas of bare soil, which undoubtedly were colonized by snowberry, bearberry, and other low shrubs, amidst which the poplar seed could fall and grow. Indeed, the spread of poplar by seed may have been much more important in pre-European times, for very large areas of soil would be exposed and maintained by animal action. Prairie dog colonies were reported by Lewis and Clark which were

11 kilometres long across the 'skirt.' The extent of bison wallows is difficult to imagine. In their work on pioneer settlement in the Canadian prairies, Mackintosh and Joerg (1934) refer to numerous surficial basins, which they termed 'blow-out soils' and which covered more than four million hectares. Accompanying photographs show that these features are very reminiscent of wallows. Possibly many of them were created or modified by the bison.

Pre-European Man and the Ecosystem

Man's place in the pre-European ecosystem was not a dominant one. Although his numbers are difficult to estimate, they were certainly far less than those of the bison, the antelope, the elk, and other major animals. Attempts to reconstruct early native populations are handicapped by the relatively small number of observations for the period prior to any expected effects from the coming of the white man. Only a few traders are known to have reached the plains prior to 1760, and it is very difficult to estimate native numbers from their journals. When European diseases began to take their toll of the natives is unknown, though a major outbreak of smallpox occurred in 1781 with several others in the early 1800s. Whether the native populations were able to build up to pre-European levels between such epidemics is not known. In spite of these problems, attempts at estimation have been made. Roe (1970) concluded that the northern plains tribes of Canada and the nearby United States probably comprised about 190,000 people in 'early days.'

Unlike modern man, these natives were very much a part of the ecosystem. In many ways they were strongly influenced by climate and other processes. For example, according to Alexander Henry the Younger, who lived on the Red Deer and North Saskatchewan rivers in the early 1800s, the natives were often compelled to follow the migrating bison herds, the principal source of meat, hides, bones and other food, shelter, and tools. On the other hand, they also influenced faunal movement and distribution. By using fire they could drive animals into traps, and by burning grass they could discourage bison from migrating into an area.

The political arrangements among the various tribes also influenced the distribution and number of bison and other animals. The plains tribes tended to occupy a loose territory which the early traders called their 'general locality,' an area in which they preferred to hunt and live. Between the general localities were buffer or neutral zones, where little or no hunting reportedly was done. The Cypress Hills area appears to have been a buffer zone separating the Blackfeet of the northern plains, the

Assiniboines of the east-central plains, and the Crows of the Yellowstone area for at least seventy years during the nineteenth century. Like other buffer zones it was rich in elk, deer, grizzly bears, and other wildlife.

The pre- and early-European natives are known to have killed very large numbers of bison and other animals. But the relatively small numbers of the Indians and low level of their technology seem to have worked against depletion by hunting. The bison were killed principally through drives, of which there were two basic types. The first, the pound, was usually a roughly circular wooden enclosure into which the animals would be driven through long lines of rocks or other 'Dead Men' leading far out into the plains. Once in the enclosure all the animals were usually slain, a practice that has been regarded as wasteful. However, apparently sound reasons have been advanced for it, among them that if bison were allowed to escape from a pound they would learn from the experience and lead their fellows away from such traps in future.

The second type of drive was the jump. Here the 'Dead Men' led to a cliff, usually eight or more metres high, over which the animals were driven to injury and death. Scores of buffalo commonly were killed in such drives. Thick beds of bone and other debris are found today at the base of some jumps; for example, the 'Old Women's Jump' in the Porcupine Hills is about twelve metres deep and was used for some two thousand years.

These drives were basic to the subsistence of the Indians of the plains so long as they were a pedestrian people. Small bands of Indians, usually kin groups, of perhaps fifty or less, would spend the winter snaring rabbits and other small animals, and running down bison, elk, and other larger mammals on snowshoes. But in the spring, summer, and fall, these groups would coalesce and carry out large combined drives in which much of the pemmican, leather, and other materials needed for the winter would be secured.

The Arrival of the White Man and Some Early Effects

With the arrival of the white man a new technology became available to the natives. Their stone tools were replaced by metal. The horse was reintroduced after an absence of some ten thousand years. It spread northward from Mexico, fitting admirably into the nomadic culture of the plains Indians. The animal could carry heavier loads than the Indian domesticate, the dog, and could move more quickly over much longer distances, greatly increasing the mobility of the native peoples and affecting their economy and life style in other ways. Its utility is reflected in the speed of its diffusion; the animal apparently was being used by

Indians only two hundred years after the Spanish brought it back to Mexico.

The horse made it less necessary to hunt in groups or drives. A few Indians could kill as many bison on horseback as hundreds of their pedestrian fellows collectively. Moreover, the mounted Indian could select the animals he wanted, notably the female for its more tender meat. To what degree the natives had such food preferences prior to the introduction of the horse it is difficult to say. Females could not be selected by using the hunting drive, except at those times of the year when the sexes grazed separately. But the horse made it possible to concentrate on the female at any time of year and so systematically reduce the breeding stock, which probably marked the first step on the road to extinction for the wild bison and the pre-European ecosystem of which it was a central part.

The strategy, attitudes, institutions, and technology of the native were in strong contrast with those of the European. By hunting and gathering they took what they needed for food, clothing, and shelter, with a certain amount of waste for technological and other reasons. They traded little with other tribes, though every year or so they might exchange pemmican or bison robes for corn or other crops at villages such as those of the Mandan on the central Missouri. The European, however, identified certain elements of the ecosystem as 'resources,' producible in large quantities for sale in growing external markets in Europe, eastern North America, and China. Principal among these resources were small fur-bearing animals such as the beaver, otter, fisher, and kit fox, which were hunted relentlessly during the eighteenth and nineteenth centuries. Moreover, the natives did much of the hunting themselves, for they were quickly incorporated into the economic system of the white man. They were attracted by alcohol, metals, beads, tobacco, horses, and guns, which outweighed any cultural barriers they might have had against the heavy hunting of furs. The respect for landscape which was inherent in the animism of some of the tribes did not stop them from engaging in excessive hunting for the fur trade. The competitive nature of the commercial economy was probably very important in this regard. For example, a late-eighteenth-century native inhabitant of the area west of Lake Winnipegosis told David Thompson that his people were hunting beaver vigorously even though they saw the animal's extinction within a few years. They were doing so because invading Iroquois or other eastern Indians would eliminate the beaver anyway.

During the late eighteenth and early nineteenth centuries competition drove the fur trade to an unusually strong assault upon the ecosystem. The

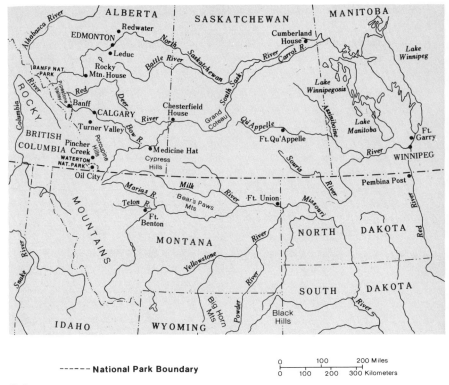

2.1

Historical Locations in the Western Interior

principal antagonists were the traders from Montreal, who eventually amalgamated to form the North West Company, and English traders of the Hudson's Bay Company. By the early 1770s both groups had established posts on the Saskatchewan River, and in the next thirty years a succession of competing houses was built westward along its north and south branches. By 1800 traders from Rocky Mountain House were draining the beaver and small furs of the foothills and mountains, and posts such as Chesterfield House, at the junction of the Red Deer and South Saskatchewan rivers, were tapping the Cypress Hills area. Such houses are said to have been capable of exhausting the beaver of a region in about seven years, although this must have involved a considerable amount of 'cream-skimming' as some continued to be productive for decades.

In 1821, after the fur companies had virtually bankrupted themselves through their competition, they amalgamated into the new Hudson's Bay Company. One of their first measures was an attempt to conserve the beaver by allowing exhausted areas to recover or 'recruit.' By 1826 a system of sustained yield was being attempted. For example, in the Saskatchewan, Lesser Slave Lake, and Assiniboine District, of which much of central and southern Alberta was part, 7800 beaver were traded in 1823, 6493 in 1824 and 6896 in 1825. The average for the three years was 7063; this was reduced by one-fifth, with the result that no more than 5561 skins were to be produced in 1826. For years thereafter, a figure of 5550 was set for the district.

However, these attempts at conservation do not seem to have worked very well in districts where growing American competition raised the same basic problems as had existed between the Hudson's Bay Company and the North West Company. The Americans had established posts on the upper Missouri about 1830 and their influence increased steadily during the 1840s and 1850s. To attract the plains tribes and drive their antagonists from the field, the Hudson's Bay Company set aside its conservational practices in those southern areas where it faced American competition. Through the 1830s to the 1850s, then, the plains animals continued to be killed by Indians pulled in two directions, north toward Rocky Mountain House, Edmonton, and other posts on the North Saskatchewan, and south toward the Missouri. Much of central and southern Alberta continued to be open to uncontrolled trapping, which probably reduced the beaver to very low levels.

The story of the virtual elimination of the bison is better known than that of the beaver, but mainly on the basis of the American example. The observations of travellers such as the Reverend Rundle suggest little evidence of depletion in the 1840s in the western Canadian plains. In 1841, he saw enormous numbers of bison in the 150 kilometres or so between the Bow and Red Deer rivers, the herds being in sight 'nearly the whole of the time.' Yet certain forces for depletion had been at work for many years, notably the long-continued pressure of the fur trade. The meat of the bison, whether fresh, dried, pounded, or as pemmican, provided most of the trader's diet, although large numbers of other animals and fish were eaten too. For example, at Fort Qu'Appelle during the late 1860s, some 1000 bison were required for subsistence yearly. During this period about fifty-three people were living at the fort, as well as thirty dogs using the equivalent of twenty men's rations.

Native people killed large numbers of bison, too, though they often seem to have been following traditional attitudes and practices. During 1883, when the bison were all but gone in a wild state, Sitting Bull led a band of Sioux 'to hunt the buffalo as in days past.' They found what seems to have been the last herd of a thousand not far from the Black Hills 'and in two days of hunting ... wiped it out to the last animal.'

The Métis or mixed bloods were an additional force for depletion. They had been increasing since the European first appeared in the west, and frequently lived around the fur trade posts. The main body settled along the Red and Assiniboine rivers, where they increased from 571 families in 1843 to 816 families, or approximately 6000 people, in 1856. Hundreds of other Métis wandered freely on the plains, visiting the settlements only occasionally. Together, the Red River and plains Métis killed large numbers of bison each year. To illustrate, in June 1840, 620 men, 650 women, and 300 children left the Red River area with 1210 carts. They are said to have returned with an average of 900 pounds of meat per cart or about 1,100,000 pounds over all. How many bison this represented is an open question, but it must have been at least 3000.

In the late 1860s and the 1870s a number of changes led to the elimination of the bison from the Canadian plains. Most important was the entrance of the American traders, who induced the plains tribes to hunt the bison for robes and hides in return for alcohol and other goods. Technological change in the 1870s led to greater demand for hides and further intense pressure on the bison. Previously, skins were collected for various purposes: carriage robes, overcoats, gloves, hats, and the like. But the leather was rather soft, and not until a new tanning process was introduced in 1871 could it be used readily for heavy-duty purposes, in competition with cowhide. Only then did bison hide become the prime objective of the hunters. The hide men operated largely in the United States, where they often killed more than 100 animals in a day and thousands in a summer hunting season. The toll is impossible to estimate with any accuracy. In 1874 the missionary George McDougall thought that more than 50,000 robes had been traded by the Blackfeet and other western plains tribes each year for a number of years and sent to Fort Benton on the Missouri. Other robes were also shipped east to Fort Garry at this time.

Working along with the cultural changes were the natural factors that had long killed large numbers of bison and held their annual rate of increase to about 18 per cent. These included predators (such as the wolf and the grizzly bear), disease, and hard winters. Prairie fires also killed large numbers, although how frequently is not clear. Rundle came across many burned carcasses while near the Red Deer River valley on 14 May

1841. Many bison also drowned each year, particularly in the spring, while trying to cross weak river ice. Such natural controls became proportionately more and more important as the herds of bison were reduced in number and confined in area, so that what would have once been a minor winter kill became a threat to the survival of the species.

In his report for 1878, the Commissioner for the North West Mounted Police stated that the best authorities were of the opinion that the bison as a means of support, even for the Indian in the southern district, would not last more than three years. In his report for the next year he commented that he had little thought that this prophecy would be so literally fulfilled. No observations of large numbers of bison are known for the southwestern Canadian plains after 1879, and very few observations of the bison were made anywhere in the Canadian plains after 1880. In commenting on its disappearance, Fuller (1961) expresses the opinion that the overriding factor was that the bison was not 'compatible with the raising of livestock or grain growing. Without the hide-hunters the bison would have held on a little longer, but only until the plains were fenced.'

The Rancher and the Farmer
According to Bennett (1969), the basic strategy of the rancher is to earn his living by raising cattle and other stock. He is not generally interested in maximizing returns but in a way of life, freedom of action, wide open spaces, and nature. His ecological posture is different from the farmer's. He uses grass, soil, and other resources *extensively* rather than intensively. Nevertheless, the early ranchers were interested in manipulating and changing nature to a much greater degree than the Indians or the fur traders. They introduced a variety of cattle, sheep, and other stock, and eliminated or severely depleted the predators as well as those animals which competed for range with their stock. At first the ranchers did little to augment or change the native grasses or forage. The animals were left on the range through the year. The emphasis was on numbers rather than on quality. As a result, as early as the 1880s there were complaints of overstocking, overgrazing, and soil erosion in southern Alberta and Saskatchewan.

The chinook is vital to the success of winter pasturing in the Canadian west, but is said to fail in roughly one year in ten. Severe winters with heavy snow and feeding difficulties must therefore be expected occasionally. It took time, however, for the rancher to learn this and to keep hay on hand as a hedge against disaster. The cold season of 1886 killed thousands of cattle and other severe winters were experienced during the next ten to fifteen years. But the prolonged, vicious, and virtually chinook-less winter of 1906 is the one most vividly remembered today by ranchers in southern Alberta and Saskatchewan. Its heavy toll convinced ranchers of the necessity to grow and keep hay.

The interest in hay, in turn, strengthened the interest in irrigation, which some ranchers and farmers had been using since the 1880s. Small dams diverted water into ditches from which it was usually spilled haphazardly over floodplains. In some places more elaborate methods were used; indeed some large ranchers like those at East End, Saskatchewan, constructed elaborate dams and irrigation systems at a cost of tens of thousands of dollars during the early 1900s. But the systems were usually simple and cheap, with little provision for the even distribution of water or for drainage. As a result, many soils became water logged or evaporation left increasing amounts of salt in the soil, to the detriment of all but the halophytic plants.

Until the 1930s the irrigation works were constructed largely at the rancher's or farmer's expense. However, with the onset of the depression and drought, the federal government began to set aside funds for dams and major diversion canals in order to provide employment, to prevent the failure of ranchers and farmers, and to 'stabilize agriculture' on the prairies. Much of this work was done through the Prairie Farm Rehabilitation Act. Although the original thrust of PFRA was to meet the water needs of the farmer threatened with failure, the agency's assistance has helped to supply increased amounts of supplemental feed and forage for many ranchers and has provided the opportunity to diversify agricultural operations in general.

Many of the difficulties faced by present-day agriculturalists reflect the views of aridity held by the members of government, the civil service, and the Canadian Pacific Railroad, all of whom were basically involved in establishing the institutional arrangements for settlement. The basic legislation was the Dominion Lands Act of 1872 which permitted a settler to claim 160 acres (about 65 hectares). Later a number of changes were made, primarily to stimulate rapid settlement. In 1874 the settler was permitted to apply for or to pre-empt an additional 160 acres adjacent to his homestead, and in 1881 group settlement was permitted, twenty families or more being allowed to form a hamlet. Settlers generally were allowed to take up the claims anywhere on the plains, no analyses having been made of the varying suitability of different areas. In the 1880s and 1890s many settlers began to experience difficulty with poor land or with drought or crop failure. Orders-in-Council were passed which permitted a settler with poor land to apply for another homestead. Seed grain and other assistance were also provided during some of the dry years.

But the government never made any basic change in the size of the land grant, even though it must have been clear by about 1900 that 320 acres of land was simply not enough on which to raise wheat and other crops profitably in many parts of the Canadian plains. Indeed, the government and the CPR are said to have initially opposed the introduction of irrigation on the grounds that it stressed aridity and might discourage settlement. Later on, the CPR, and eventually the government, vigorously promoted irrigation and it was often the farmers who failed to use it consistently. They permitted irrigation works to fall into disuse in relatively wet periods and repaired and elaborated upon them in dry periods.

The farmer has been much more instrumental in changing the ecosystem than the rancher. From the beginning his strong orientation was to monoculture, to growing a single crop such as wheat over large areas of former grassland. An impression of the effort involved can be derived from the fact that the fescue plant community embraces about 148 species of higher plants, of which over 70 occur consistently in areas several hectares in size. Not unnaturally then, 'weeds' were a major problem for farmers. The earliest means they seized upon to keep the native plants out of the fields was cultivation. Horse-, steam-, and automobile-driven plows were used to pulverize the soil into a fine dust from which as much vegetable life as possible was extinguished. This dry mulch also was also thought to be a useful means of promoting water absorption, although the splash of raindrops soon closed the fine surficial soil pores, promoting sheet wash and water erosion. When dry, the level dust mulch also was very susceptible to the wind erosion that became so noticeable in the late 1920s and during the 1930s. As these problems became severe, particularly during the depression, other means of cultivation were invented, such as the duck-foot plough which could break the soil some inches beneath the surface, killing the weeds and promoting less soil erosion. Other dry farming and soil conservation techniques, such as stubble mulching and terracing, were also introduced, and chemicals began to be used to control weeds, though their heyday was not to come until after World War II.

Rodents, insects, and fauna caused problems for the farmers, who had difficulty tolerating any animal that might damage their crops. Large-scale campaigns were waged against the small fauna in the late nineteenth and early twentieth centuries. School children were encouraged to hunt ground squirrels and gophers during the summer and prizes were awarded for the largest kills. It was not uncommon for the winners to turn in thousands of tails.

Locust outbreaks were a major cause of crop losses up to the 1930s.

Various poisons were developed to kill these insects, but there was more difficulty with rusts and other infestations of wheat and grains. Thus the wheat sawfly could be controlled in the fields, but would shelter in the winter in the native wheat grasses of the surrounding country. This led to attempts to remove these grasses from ditches and roadsides and to replace them with poorer hosts, such as an exotic brome grass from Europe. Its heavy golden spike is now a common sight along Alberta roadsides, along with Russian wild rye and other imports. Unlike the native grasses, early maturing exotics such as the wild rye can withstand heavy grazing in the spring and early summer. They are still used widely as early forage over much of the prairies.

Bennett (1969) is one of the few scholars who has compared the attitudes of western Canadian farmers and ranchers to nature, on the basis of his studies in the Cypress Hills area. He found that the rancher considers himself as part of 'unspoiled' wild nature, a partner with the coyote, the grass, the coulees, and the open spaces – however much he may in practice have to modify that approach. The farmer, on the other hand, sees nature as a wilderness in need of taming. Thus, though the rancher might be proud of the elk or antelope that eat his hay in winter (complaints to the game warden notwithstanding), the farmer generally sees wild species as pests which eat grain, tear up fields, and kill chickens. He does his best to eradicate them. At the same time, the farmer has another kind of respect for nature, a respect born of competition. In the Great Plains, in particular, nature is fraught with unpredictable hazards – the late frost, the failure of spring snow or of the winter chinook, summer hailstorms, tornadoes, droughts, windstorms, and floods. In learning to deal with these things through dry farming and in other ways, in attempting to tame or conquer them, the farmer has caused many unfortunate changes in the landscape and the ecosystem, but at the same time, in Bennett's opinion, he has developed a profound respect for nature and much better conservation farming practices.

Bennett also says that ranchers and farmers, as well as many plains urban dwellers, have made symbolic adaptation to the landscape. They have created conditions which make it easier to adapt mentally to the environment. Many settlers were originally from the east, which may in part explain why their aesthetic preferences, and those of their offspring, are centred on the idea of a green and moist land. They display much interest in 'natural' beauty, which is often defined as the 'greenery' in flower gardens or tree belts. The search for the 'green' goes to considerable lengths, as witness the multitudes of gardens and lawns in the western cities at great cost in terms of water and money.

Public Lands, Especially National Parks

The grassland and parkland are by no means alone in experiencing the heavy hand of man. The provincial forests, for example, were set aside following extensive lightning and man-caused fires during the late nineteenth and early twentieth centuries. The heavy destruction of timber raised questions about the ability of the foothills and the highland watersheds to supply the steady flow of water thought necessary for agriculture and settlement on the dry plains. Reserves were therefore set up by the federal government in 1911, and transferred to the control of the provinces in 1930. Even the mountain national parks are not 'untouched' landscapes. They were trapped, prospected, mined, lumbered, transected by railroads and other routeways, and burned in much the same manner as most of western Canada in pioneer days. The principal impact in the Banff area, for instance, followed the introduction of the railway in the 1880s. Surveyors caused a number of fires in the Bow Valley country. The sparks thrown by early wood- and coal-burning engines caused many more in the next several decades. Only a few large areas, such as the Upper Red Deer Valley, escaped these conflagrations and remain largely covered with the pre-European climax forest of spruce and fir. Much of the rest of the park is clothed with even-age, fire-following lodgepole pine stands which have developed under increasingly strict fire control policies since about 1911.

The fauna of the national parks has also changed considerably under the hand of the white man. David Thompson travelled up the Red Deer River to the vicinity of the present boundary of Banff National Park in the fall of 1800 and observed many bison, elk, moose, and small fur bearers such as the fisher. Later travellers, such as James Hector in the late 1850s, also saw deer, mountain sheep, goats, and other wildlife in various parts of the foothills and Banff park area. By the 1880s the Mounted Police were complaining about the high toll that railway men, miners, Indians, and others were taking of fauna. In 1886, the early settlements on the eastern slopes of the Rockies were said to be surrounded by belts of country, perhaps twenty-five miles wide, in which all forms of 'big game' had become extinct.

Hunting was carried out as a sport and recreation for a number of years after the establishment of Rocky Mountain Park (later Banff National Park) in 1887. Government policy also favoured the reduction of predators and so-called noxious animals. An early wildlife study recommended that 'wolves, coyotes, foxes, lynxes, skunks, weasels, wildcats, porcupines and other animals be destroyed.' Protection did increase after 1910, however, and a varied and numerous wildlife population can now be found in the national parks, although the population differs in kind and number from that first observed by the white man. The bison is gone and the wolf is very rare, but elk are quite numerous and place such heavy pressure on the vegetation that their numbers have to be controlled.

Power, Especially Oil and Gas, and Urbanization

Power production has been the basis of much of the recent economic and population growth in the prairie provinces and has had considerable impact on the landscape. The mining of coal, the damming of rivers, and the development of oil and natural gas are all involved. Many abandoned open-pit mines can be found already and more are likely to be produced, particularly along the mountain front. Hydroelectric projects have been under construction in the area since the late nineteenth century, and some rivers (e.g., the Bow) are now well endowed with dams and power plants. Many, however, have not been affected yet and careful consideration should be given to their preservation as wild or scenic rivers.

Oil was known to the fur traders, who observed it along rivers such as the Athabasca as early as the 1780s and 1790s. A commercial well was drilled in the Medicine Hat area in 1890, but the first major find was the Turner Valley field in 1914. During the next twenty years about two hundred wells were drilled in this area in uncontrolled fashion. Large quantities of gas were released in drilling and production. There was no planning of the location and spacing of wells, or of production or marketing. Many producers attempted to get as much oil to market as quickly as possible; prices therefore were low, stability of operations difficult, and conservational practices poor. The first of the major post-war discoveries was made at Leduc, near Edmonton, in 1947. About twenty months later, the nearby Redwater field was found, and by 1951 it was producing 23.2 billion barrels a year. Thereafter came a stream of discoveries which have continued through the 1960s and have extended the oil industry beyond Alberta.

All this activity has brought a number of environmental problems. In Alberta, these are handled by the Oil and Gas Conservation Board, which was established in 1938 because of the marketing difficulties and waste in the Turner Valley field. Over the years, the Board has established a number of operational and conservational policies, such as setting a minimum space of about 16 hectares around each oil well. The Board also worked out a system of maximum production rates for the Turner Valley field, but this ceased to be meaningful after the Leduc discovery. By 1950 only 66 per cent of the Alberta production potential could be marketed and that percentage continued to fall for the next few years. The production companies attempted to set their own quotas but eventually the Oil and Gas Conservation Board had to establish a pro-rating system. In recent years, the Board has become interested in enhancing the recovery of oil and gas, and it approved 43 schemes in 1969. Over 90 per cent of these included water flooding, with the gain in recovery expected to approach two hundred million barrels. The Board also is heavily involved in attempting to control water and air pollution. For example, if an operator wishes to drill a well near a freshwater body he must first gain the Board's approval. The plans must provide for the construction and maintenance of dikes, reservoirs, or other installations to contain any spills which might pollute the water. The method for disposing of mud, oil, or other wastes from drilling must also be shown. These conservation measures seem generally to have been effective, though there have been occasional problems. For example, there have been serious complaints of air and water pollution in the Pincher Creek area of southwestern Alberta.

The activities of the oil and gas industry have also caused other types of environmental change. Seismic trails are a special problem. The banks of a stream may be broken down, to the detriment of fishing, and serious local erosion may follow the cutting of seismic lines. Where oil exploration is extensive, large amounts of salt water can also be added to streams.

But the greatest effect of the oil industry on the environment is an indirect one, arising from its basic role in promoting very rapid economic development and population growth in Alberta. Much of this economic development has been beneficial, but the population growth is concentrated in the cities. For example, the population of Calgary rose from 125,000 in 1951 to 400,000 in 1971. This urban growth has been accompanied by many environmental and social problems usually associated with larger cities. Thus, the sewage which until a few years ago enriched the Bow River and its trout fishery, now exerts a very high bio-oxygen demand and contributes to pollution along with the growing quantity of toxic and non-toxic wastes from the refineries and other industries in the city. Air pollution is also becoming an increasing problem in the cities and towns of the prairies, notably those subject to chinooks and other causes of the temperature inversions which hold sulphur dioxide and other irritants near the ground. Complaints are also being made about noise pollution and the aesthetic and recreational inadequacies of the urban environment. Recent studies suggest that many residents of Calgary leave the city only two or three week-ends a summer. Some of this may be due to chance, but other people are held in by low incomes, old age, or other problems.

To correct these problems will require more changes in strategy, institutional arrangements, technology, and the ecology of cities and their hinterlands. Suggestions are increasingly being made that population growth should be stopped, city sizes controlled, and smaller satellities created in more pleasant natural environments some distance from the city and connected to it by some form of rapid transit. Care would have to be taken, however, that such changes do not merely spread the problem for a time, rather than reduce or eliminate it.

The solutions to the burgeoning landscape changes in the prairie provinces seem chiefly to lie in the control of population growth and technology and in effecting better institutional arrangements for the management of resources and the ecosystem. A maze of governments and agencies is involved in the use of water and other elements of the landscape, and insufficient co-ordination exists among them. One recent suggestion is that Alberta, Saskatchewan, and Manitoba should be amalgamated into a single prairie province. In many ways this would be a forward step, but there are other, more easily realizable needs, among them environmental councils and the input of more ecological information in decision-making processes. The recent move by the Alberta government to establish an Environment Conservation Authority, with scientific and public advisory councils, is an encouraging portent.

3 Agriculture

BRUCE PROUDFOOT

Agriculture has played a dominant role in the political and economic development of the prairie provinces. To many people, both in Canada and abroad, the very term 'prairie' is synonymous with vast fields of golden grain, rather than with the original grassland vegetation of a small part of the three provinces. Yet agriculture, of any sort, occupies less than half their total area and, today, contributes less than one-third of the value added in the goods-producing industries and employs, directly, fewer than one in seven of the total labour force. The change in the prairie provinces from a dominantly agrarian economy is relatively recent, and the social and economic problems of rural decline and migration to the major urban centres have given rise to current political concerns in each of the provinces. The role of prairie agriculture is equally a matter of political concern at the federal level, for the traditional importance of prairie wheat exports in Canadian foreign trade has declined, and marketing and transportation have been influenced or controlled by federal legislation. Moreover, it has been difficult and costly for western livestock and feed grains to penetrate eastern Canadian markets in the face of strong competition from producers nearer those markets. The cost of distance added to the hazards of a difficult environment are of basic importance in the agriculture of the prairie provinces.

Although the first truly agricultural settlement in the provinces was as early as 1812, with the establishment of the Selkirk colony in the Red River Valley, the colonization of the area for agriculture began in earnest only towards the end of the nineteenth century. There were many reasons for this. Competition between the Hudson's Bay Company and its rivals led to uncertainty over surety of tenure, so that there was unwillingness on the part of individual entrepreneurs to commit themselves to the kind of long-term investment demanded by agriculture. This particular problem was solved finally with the purchase of Rupert's Land from the Hudson's Bay Company in 1869, and the development of government policy for settling the area. It was not, however, this political change that alone permitted occupation of the vast western lands by agricultural settlers. Nor was it solely the railway which was pushed westwards to link the youthful nation of Canada, founded only in 1867, with the far-western province of British Columbia, which joined Confederation in 1871. Certainly the railway provided the transportation links by which many new settlers arrived, and by which most of the prairie's agricultural produce left the region, but other factors were also of importance. As the rail links were being built, the North West Mounted Police took effective possession of, and asserted Canadian sovereignty over, the territory lying north of the 49th parallel. Concurrently, beyond the prairies, in eastern Canada, in the United States, and in Europe, especially in Britain, the first large-scale attempts were being made to recruit settlers for the agricultural colonization of the west. Such colonization was ultimately feasible only because of developments in agriculture itself, namely, the production of early maturing varieties of wheat which could ripen in the short summer season that was characteristic of northern continental plains. The old Red River variety of wheat was replaced by Red Fife, which remained dominant on the prairies until 1920. Another variety, Ladoga, ripened four days earlier; and a third, Hard Red Calcutta, matured ahead of Ladoga. Marquis wheat, produced by crossing Hard Red Calcutta and Red Fife, ripened some eight days earlier than Red Fife. It gave a heavier yield, was more resistant to drought than some other varieties, and had good milling qualities. After its introduction into the west, from the Dominion Experimental Farm in Ottawa where it was produced, just prior to World War I, it rapidly spread and replaced Red Fife as the dominant variety. The windmill pump, the barbed-wire fence, the self-cleaning steel plough, and the self-binding reaper, developed and used in the American Midwest and Great Plains, provided further technical inputs which made the settlement of the Canadian prairies feasible. From an early stage in the settling of the area, farming was considerably more mechanized than in other areas, especially those in Europe from which many of the new colonists migrated. This factor, combined with available overseas markets, made accessible by railway and steamship, determined that prairie agriculture was from the start commercially oriented. Subsistence agriculture was a temporary phase in the more remote areas, before these too were linked by the railroad to the markets.

Settlement was slow even after the completion of the transcontinental railway in 1886 as settlers probed the difficulties of the new environments in which they were settling. Experimental farms were established by both the federal government and the Canadian Pacific Railway Company to provide answers to farming problems and promote better farm management. Late in the 1880s, at the Dominion Experimental Farm at Indian Head in Saskatchewan, the practice of black summer-fallowing was developed, whereby the fallow land was cultivated during the summer to reduce weeds and prevent the loss of moisture through evaporation, so that increased supplies would be available the following year. Perhaps the greatest contribution made by the railway company was the development, after 1903, of irrigation techniques in southern Alberta, although the company was also influential in promoting higher standards of farming over a much wider area than that which it owned.

Economic circumstances beyond the prairies, as well as advances in farming techniques, and a period of greater and more dependable rainfall, increased the rate of settlement towards the end of the nineteenth century. Transport costs were reduced; for example, charges on wheat from Regina to Liverpool dropped from 35 cents to 21 cents per bushel between 1886 and 1906. After 1897 the long decline in wheat prices ended and the downward trend was reversed. The American West had filled up, land prices were rising, and free land in Canada was attractive to those younger sons who would not inherit the family farm. As population increased and the cultivated area expanded, so too did the railway net, integrating more closely the economy of the prairies with external markets. The expansion in agriculture is indicated by the increase in the area of principal grain crops from 600,000 hectares in 1891, to 1.0 million in 1898, to 3.8 million in 1908. By 1901 there were 419,512 people in the west. By 1911 this had more than tripled to 1,328,121, while the area under all field crops had increased in the same period from 1.5 million to 7.2 million hectares. As farmers settled the empty lands a mesh of service centres developed, especially around the grain elevators along the railway lines. Grain elevators were built about every thirteen kilometres along the railway and the numerous small centres provided services for the farm population living within a radius of about sixteen kilometres, for that seemed to be the economic limit for grain delivery by horse-drawn wagon. Interestingly, the mesh of smaller centres was best developed in Saskatchewan, which has been, throughout this century, the most rural of the three provinces, but even in that province by 1911 nearly 30 per cent of the population was classed by the Dominion Census as non-rural.

Most of the farm settlement was based on the township and range system which was established in 1872 with the proclamation of public lands in the west, under the Dominion Lands Act. Under this legislation, a settler could take up a quarter-section (that is, a half-mile square unit of 160 acres, about 65 hectares) of unoccupied land in the prairies and work it as a 'free homestead.' Payment of a 10-dollar registration fee was required, and after three years the homesteader could file a claim of ownership providing he had improved his property and resided on it for specified periods during that time. In some areas it soon became apparent that a quarter-section was insufficient to support a family and under a 'pre-emption' system, which remained in effect until World War I, settlers were permitted to purchase an adjacent quarter-section at the nominal price of one dollar per acre. Apart from the public lands which were essentially free to settlers, the Hudson's Bay Company retained one-twentieth of the land south of the North Saskatchewan River and one-eighteenth of all land was set aside as school lands. In 1880, 10 million hectares were allocated towards the cost of the transcontinental railway, and at varying dates after this some 1.6 million hectares were similarly allocated to other rail lines which were subsequently acquired by the Canadian Pacific. Such land was excluded from settlement except by purchase, or by lease. Although the alternation of sections of Crown and railway land caused difficulties in some areas, the use of its lands by the Canadian Pacific as part of company policy to foster the buildup of settlement, and thereby generate traffic for the railway, was of considerable value and importance. Through its policies the company enabled many farmers to sit out the collapse of the land boom in 1882, the poor crops in the years that followed, and the low grain prices prior to 1897.

Quarter-sections were laid out irrespective of physical conditions for agriculture. At the time of settlement little information was available on such factors as climate and suitability of soils, and the information which was available was often contradictory. The haste with which settlement was carried out did not allow time for such considerations, and the farm frontier expanded at enormous cost in human effort. Some estimates suggest that for every successful settler who managed to establish a viable farm enterprise in the northern part of the prairies there was at least one other who failed. In areas of aspen woodland in west-central Alberta, some quarter-sections which were physically marginal for any sort of agriculture were homesteaded and abandoned as many as six times. Optimism and doggedness were necessary qualities for pioneering. Survival and adaptation to the environment were linked, and some have seen in this conflict the origins of the independence and self-reliance of the westerner not only as an individual, but collectively in his political activities, which have differed sharply from those of the rest of Canada and have continually been reflected in the divergent views of the prairie and federal governments. Many of the techniques brought into the region from outside were of little value until modified to suit the hazardous environment. The problems varied between locations, and during the year. To simplify, in the drier, southern areas the problem was rain – its amount and frequency. Farther north there was the problem of frost, both in the spring when a late frost could kill the young crop, and in the late summer when an early frost could damage unripened grain. Hail provided another climatic hazard, and grasshoppers, sawflies and rust were biological hazards which could decimate the growing crop. Few of the settlers entering the west for the first time had any real farming experience under similar kinds of conditions. There were some exceptions, particularly in the case of the Mennonites who had had experience of similar physical environmental conditions in the steppe lands of southern Russia, before settling in Manitoba in 1874. Americans from western rural areas immediately south of the international frontier also had experience of a similar physical environment and a similar settlement system based on individual ownership of family farms within the framework of the township and range system. The experiences of the successful farmers were used widely in the promotional literature designed to attract further settlers.

In spite of their shortcomings, homestead programmes enabled individual settlers to become established in farming with a minimum amount of capital. The fact that the homesteader could, aided by his family, cut wood for fuel, grow feed for his livestock, and perform most of the routine farm labour himself also assisted in keeping his money or capital needs to a minimum. He could, to varying extents, obtain a supplementary meat supply by hunting and, especially in the better-wooded areas of the parkland belt, fell timber for housing. Part-time, off-farm employment was also available, for example, during the construction of the railroads, and more generally in sawmills, lumber-yards, and such other industries as construction. As the productivity of the land was developed, its worth increased and the values so created accrued to the owner. Hence the land grant system provided agriculture with a large part of its original capital input. It should not be forgotten, however, that there were capital inflows into the region during the period of settlement. Although many settlers had individually limited amounts of capital, in total the amount entering the area was considerable. Then, too, there was the social capital represented, for example, by the earlier education of the original settlers themselves, a capital cost borne by the countries from which the settlers had come. Some estimate of this can be obtained by the very costs which the settlers had to bear in providing similar services in education, roads, and churches in a previously unsettled land. The railways also represented a massive influx of outside capital, for railway construction was financed very largely from outside Canada, and certainly from outside the prairie provinces. Financial inflows from outside the region were also provided by the railways in a number of other ways. During the construction period and on occasions afterwards when rail costs were greater than receipts, employment by the companies of settlers on a part-time basis represented external inputs into the local agricultural economy; so also did subsidized fares, reduced rates of interest on land payments, and technical advisory work in agriculture.

Within the constraints of the homestead system, increased knowledge of the physical limitations of the environment, and of the means to modify agriculture, enabled adaptation to take place. This has been a continuous process, concomitant with the expansion of the continuously settled area to its present limits, which are determined not by physical factors alone but also by the price of the products harvested in relation to the costs of production. South of these northern limits of agriculture and east of the limits of continuous settlement where the plains meet the foothills, there are few areas which have not been cultivated, the most important of which are in the drier areas of southern Alberta and southwestern Saskatchewan where ranching has been, and in some areas still is, an important land use. Prior to the spread of farm settlement westwards, ranching developed in the drier areas of southern Alberta largely as an extension of ranching from Montana. The Dominion Lands Act of 1872 permitted settlers to obtain grazing rights to land adjacent to their homesteads. Grazing leases were permitted to anyone in 1876, and in 1882 twenty-one-year leases of areas up to 4000 hectares in extent were allowed. If such lands were needed for homesteads, leases could be cancelled on two years' notice.

Initially, ranching involved an extremely extensive use of open and free range, but with pressure on the land from homesteaders after 1882, and a substantial increase in the number of ranchers themselves, there developed a more intensive use of leased and owned land. In the early 1880s ranching was concentrated especially in the Rocky Mountain foothills of Alberta but later spread eastwards onto the plains. As pressures on the land base increased, large commercial ranches gave way to owner-operated small ranches and mixed farms. Especially after the turn of the century some of the larger ranchers moved farther eastwards into the driest areas of Alberta and southwestern Saskatchewan which had not been previously settled. However, almost as soon as they had begun operations they were again under pressure from settlers using continually improving dry farming techniques. The ranchers were forced to buy land and operate in much the same way as other settled farmers, so that by World War I large-scale ranching on open and leased range, comparable to that in the western states of the United States, had virtually ceased in this part of Canada. However, ranching on a smaller scale has remained as a viable enterprise in the foothills area of southern Alberta, its economic activities geared closely to the annual climatic cycle which includes the vagaries of late spring frosts, summer droughts, early fall frosts, and winter chinooks and blizzards. Wheat, barley, and forage crops are grown to varying extents by the ranchers themselves and on farms on lower areas adjacent to the ranches. Young feeder stock is sold from some ranches, whereas others retain and fatten their steers. Through the whole of southern Alberta, and to varying extents elsewhere in the cattle-rearing areas of the prairies, seasonal stock movements occur to take advantage of lateral and horizontal variations in available grazing. Outside the dry southern areas large-scale cattle operations using extensive patterns of grazing occur along some of the settled margins of agriculture, but these developments have not been on the scale found in the British Columbia section of the Peace River country where different provincial legislation has permitted the development of large corporate enterprises. Within the prairie provinces current pressures on ranching lands come not from farmers, as has historically been the case, but from such non-agricultural activities as recreation, conservation, and mineral extraction. The continued viability of ranching as a means of rearing and fattening livestock in competition with intensive feed-lot operations is likely to be a problem in the immediate future.

Such intensive feed-lot operations are among the most recent changes that have occurred in prairie agriculture in response to technological innovations and changing economic circumstances. From the very initiation of settlement the impact of technology has been clear, and it was the same technological developments that had enabled settlement to occur which, virtually from the start of homesteading, caused some of the problems associated with the quarter-section farm. Whereas 160 acres had earlier, and in more favourable physical environments, enabled a farm family to support itself and produce a modest surplus, such an area in the prairies was insufficient to produce the volume of grain needed to support a commercialized and mechanized agriculture so remote from markets that transport costs were inevitably considerable. As a result, the average size of farms increased, slowly until the mid-1940s, and rapidly thereafter as the scale of mechanization altered with the disappearance of horse-drawn equipment. Increased mechanization, especially after World War II, enabled the individual operator to work a much larger acreage than before, and the disappearance of horses freed further areas for commercial grain production.

The result of these changes is that prairie agriculture, like that of much of the rest of Canada, is now a capital-intensive industry. Investment in the basic resource – land – remains important. However, especially within

Table 3.1   Capital value of farms by province, 1969

	Manitoba	Saskatchewan	Alberta
Land and buildings	65%	68%	67%
Implements and machinery	23%	23%	19%
Livestock and poultry	12%	9%	14%

Source: Based on data from Canada Department of Agriculture, Canadian Imperial Bank of Commerce *Commercial Letter*, Nov./Dec. 1971, p. 3.

the last 25 years, emphasis has been placed on investment in buildings, machinery, equipment, and the application of other improved technological innovations which enable the individual worker to handle larger acreages and greater numbers of livestock more efficiently. Labour, which even as late as the mid-1940s represented over half of the total input in Canadian agricultural production, had dropped to 35 per cent by the early 1960s, while machinery and equipment at 22 per cent were almost equal in importance to real estate at 24 per cent. All of the remaining inputs such as feed, seed, and fertilizers have increased in relative importance. More significantly, from the point of view of the farmer himself, the change in inputs has involved much larger, regular, cash payments. As inputs have increased in cost there has been generally no corresponding increase in the price the farmer has received for his products. In some cases, prices have tended to decline; for example, the price, per bushel, of No. 1 Northern Wheat, at the Lakehead, in the middle of 1971 was some 25 cents lower than it was in the middle of 1964.

An outstanding feature of prairie agriculture, and of Canadian agriculture in general, is the very great difference in capital structure between the large and small farm units. Provincial average capital investment per farm ranges from a low of about $26,000 in the Atlantic Provinces to a high of $75,000 in Alberta. The average breakdown of capital value of farms by province (Table 3.1) indicates little variation among the prairie provinces. The overwhelming importance of the value of land and buildings is clear. The variation in the relative value of livestock and poultry reflects the minor role of this sector of farming in Saskatchewan generally. However, the averages obscure the situation with regard to specialized farming operations, and regional variations in the agricultural structure within each province. Farmers owning large, highly productive grain farms in the southern prairies are likely to have capital investments of $200,000 and over, as are cattle ranchers in the Alberta foothills. In contrast, the capital value of a newly cleared homestead in northern Alberta or a small farm along the northern edge of cultivation with less than 80 hectares under cultivation is unlikely to be as high as $20,000.

The largest farms are found in the drier southern regions of the prairies, and, in general, farm size decreases northwards towards the settlement frontier. Two major anomalies to this general pattern are the regions of small farms in eastern Manitoba and in the central part of southern Alberta, between the large farms of the dry Palliser triangle and the foothill ranches: some farms in this second group are irrigated. Farm size is not simply a reflection of variation in climate, or in soils. It is also a reflection of the age of settlement. Most farm sales, or transfers in ownership, occur when farmers reach retiring age, and in the older settled areas there have been more opportunities for such intergenerational changes to occur. Moreover, it was in the drier, first-settled, areas that drought was first recognized as a hazard in the 1880s, and it was in these same areas that the impacts of lower cereal prices, drought, and wind erosion were most severely felt in the 1930s. Considerable farm abandonment took place at that time, followed by the relocation of the settlers along the northern margins of settlement, especially in the Peace River country. Changes in farm size have also been brought about by government intervention. In 1938 large areas near the Red Deer River in Alberta were designated as Special Areas unsuitable for agricultural settlement, and were withdrawn from cultivation by the transfer of privately owned land to the Crown, some of which was converted to grazing leases. Similarly, in neighbouring areas of Saskatchewan, community pastures were set up under the auspices of the Prairie Farm Rehabilitation Administration. An institutional framework was thus created to cope with the environmental and economic problems of the times. From an agronomic point of view, technical developments, notably at the federal research stations at Swift Current in Saskatchewan and Lethbridge in Alberta, and in the provincial universities, were of equal importance. Strains of crested wheat grass were developed to assist in the control of drifting soil, and over a hundred thousand hectares were recovered between 1937 and 1941. Subsurface cultivation, stubble mulching, or trash farming, and alternate strip cropping, were all developed to conserve moisture, inhibit erosion, and control weeds. Finally, the development of wheat strains more resistant to rust than Marquis reduced another hazard. Collectively, these advances in technology enabled agricultural adjustments to occur after the dry erosive years of the 1930s on a scale scarcely less significant than the original settlement of the area, which had, likewise, been made possible by technological advances in agriculture.

By 1951 substantial changes in the patterns of crop distribution had occurred. Wheat growing had decreased considerably in importance in central Alberta, western Saskatchewan, and southwestern Manitoba. Increases in wheat growing, sufficient to counterbalance these, had

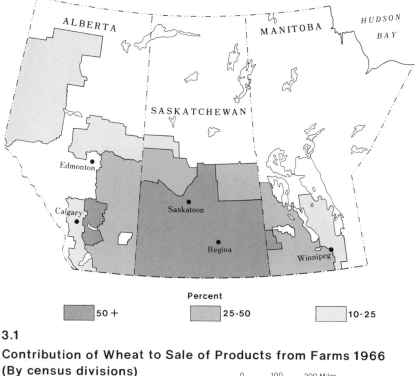

Percent

50 +   25-50   10-25

3.1

**Contribution of Wheat to Sale of Products from Farms 1966
(By census divisions)**
(Based on Census of Canada Data)

0   100   200 Miles
0   100   200   300 Kilometers

occurred in southern Alberta, central Saskatchewan, and the northern
marginal areas of all three provinces. Central Saskatchewan and southern
Alberta have retained their importance as wheat-growing areas until the
present, although there have been shifts in the relative significance of
wheat in the different provinces, and regionally within the provinces
(Figure 3.1). In some areas there has been a significant reduction in the
acreage of wheat on the poorer classes of land and substitution by other
cereals or pasture for which the areas are physically better suited. Over
all, the acreage of cropland has continued to increase, with a present total
of approximately 30 million hectares in principal field crops and summer-
fallow. Wheat acreage reached its recent maximum of 12 million hectares
in 1967, declined slightly in 1968 and 1969, and was halved in 1970 to
some 4.9 million hectares as a result of the federal government's LIFT
programme (Low Inventories for Tomorrow) which encouraged farmers
not to plant wheat but to replace it, in the short term, by summer-fallow

Table 3.2   Area under wheat, 1969 to 1971 (million hectares)

	1969	1970	1971
Manitoba	1.0	0.6	1.0
Saskatchewan	6.7	3.2	5.2
Alberta	2.1	1.1	1.4
Total	9.8	4.9	7.6

Source: Canada Department of Agriculture, *The Canadian Agricultural Outlook*, annual publications for 1971 and 1972.

or, in the long term, by other grains or forage crops. The proportional decline in wheat acreage within the last two years has been slightly greater in Alberta than in the other provinces (Table 3.2). Total acreage under wheat increased to some 7.6 million hectares in 1971, the main increase being in Saskatchewan and the smallest in Alberta.

The role of wheat in the provincial agricultural economies differs significantly. In the crop year 1965–6, for example, some 65 per cent of total farm sales in Saskatchewan were derived from wheat as compared with 25 per cent for Alberta and 31 per cent for Manitoba. The importance of agriculture as a whole differed significantly among the provinces, being greatest in Saskatchewan. The combination of these two sets of relationships accounts for the varied contribution of wheat to the Gross Provincial Income, and hence the differential impact of fluctuating wheat receipts on the provincial economies (Table 3.3).

Agriculturally the most diversified of the provinces is Alberta. This is in part the result of the decline of wheat growing in central Alberta since the 1930s and its replacement by forage crops and especially by barley, which economically competes strongly with wheat. Other diversification provincially comes from the emphasis on livestock through the whole of the western part of the continuously settled area – in the southwest of the province a continuation of the ranching which was the first agricultural use of the area (Figure 3.2). Local specialization to varying extents also contributes to provincial diversification: sheep in parts of southern Alberta; potatoes, vegetables, and sugar beet, especially in the areas east of Lethbridge; dairying in central Alberta; rapeseed along the northern margins of agriculture; rapeseed, grass seed, and other small seeds in the Peace River country.

Diversification in Manitoba has been of longer standing than in either of the other provinces. Already by 1921 less than half of the acreage under field crops was occupied by wheat. Potatoes, vegetables, and sugar beet are especially important in the areas east of Morden and between

Table 3.3   Role of agriculture and wheat production, by provinces, 1961–6

	Value of wheat sales as % of total farm sales 1965–6[1]	Percentage occupied in agriculture 1961[2]	Percentage value added in goods-producing industries by agriculture 1962[3]	Wheat share as % of Gross Provincial Income 1961–4[4]
Manitoba	31	18	29	3
Saskatchewan	65	34	55	15
Alberta	25	21	25	7

1   Dominion Bureau of Statistics, *Census of Canada*, 1966, vols. 5–1, 2, 3 *Agriculture*, Table 23.
2   Dominion Bureau of Statistics, *Census of Canada*, 1961, vol. 3–2, *Labour*, Table 1.
3   A.L. Boykiw, 'One Prairie Province and Regional Development,' in D.K. Elton (ed.), *Proceedings of One Prairie Province? A Question for Canada and Selected Papers* (Lethbridge, 1970), p. 385.
4   Hedlin-Menzies Ltd., cited by H.B. Huff, 'Implications and Alternatives of the Wheat Market Outlook,' *Canadian Agricultural Outlook Conference* 1969 (Canada Department of Agriculture, Ottawa), vol. 2, p. 76.

Winnipeg and Lake Winnipeg along the Red River valley, the area first settled for agriculture in the west. Scattered through southern Manitoba, corn for silage is very much more important than elsewhere in the prairies. The province, like much of Alberta, is dominated regionally by mixed farming.

Saskatchewan is agriculturally the most specialized of the provinces. Wheat has dominated the economy, and other cereal crops, such as barley and oats, have been much less significant than in the other provinces. During the early 1960s the wheat acreage in central Saskatchewan increased at the expense of oats and flaxseed, while the area of summer-fallow increased as many farmers followed the injunctions of agrologists to make fuller and better use of their land than is possible under a simply alternating wheat-fallow rotation. Such changes were most marked on the better soils so that production figures for wheat increased even more than did the acreage under crop. Specialization is greatest in the central part of the province where wheat sales amounted to 75 per cent or more of the total value of farm sales in the mid-1960s. In a broad belt about 160 kilometres wide along the northern limits of settlement and in the east of the province, grain crops other than wheat and mixed farming are more important. Especially in the northern area rapeseed production is widespread, but the acreage has increased in areas farther south within the last few years in response to improved international markets.

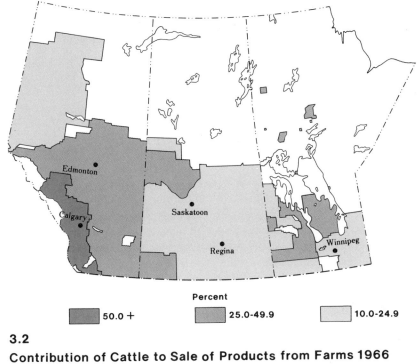

Percent

| 50.0 + | 25.0-49.9 | 10.0-24.9 |

3.2

Contribution of Cattle to Sale of Products from Farms 1966 (By census divisions)

(Based on Census of Canada Data)

Recent changes in crop production serve in many ways to emphasize the current dilemma of prairie agriculture. Increasingly sophisticated methods of production are available which will increase yields, yet, concurrently, the market for many prairie products is declining or at best static, often in terms of both volume demanded and price offered. Moreover, the major markets for prairie products are external to the region, if not outside Canada. Only in limited parts of the Canadian market, and nowhere on international markets, do prairie producers command a dominant or near monopolistic position, so that they generally have little impact on market prices. Equally, they have little control over the costs of many of their inputs, such as farm machinery, for the prairie market is rarely the only market in Canada for such items and transport costs are inevitably high because of the wide dispersal of the small farm population. In this context, the relative importance of Ontario in Canadian agricultural production should be noted.

In the early 1960s the value of farm sales in Ontario amounted to approximately half that in the three prairie provinces. In 1971 federally inspected slaughterings of cattle in Ontario were still more than half those of the three prairie provinces in spite of a substantial increase in slaughterings in Alberta. Similarly, feed grain production in Ontario is still one-third that of the prairies, even with massive shifts from wheat production into feed grains in the prairies during the last few years.

The prairie producers and those in Ontario especially, but others also in eastern Canada, are competing in some of the same markets. Political events relevant to agriculture late in 1971 also reflect the competing interests within Canadian agriculture, and the difficulties of creating a national agricultural policy which is acceptable to the diversity of producers throughout the country. The acrimonious debates over federal bills to establish farm stabilization programmes, and produce marketing boards, brought out clearly the differences in problems and attitudes between the western producers and other Canadian farmers. Overseas markets in oil-seeds and feed grains have been at record highs for Canada in the last two years, and have compensated the prairie growers, to some extent at least, for the decline in wheat markets. Any contraction in these overseas markets will certainly increase the internal conflicts within Canadian agriculture. Whatever the strength of the external markets, prairie agriculture is likely to face a difficult period of further adjustment. The conflicts between gearing production in a hazardous physical environment to fluctuating markets, and at the same time providing farmers with an adequate return on their labour and the capital invested in their farms, are still far from resolution.

Within agriculture as an industry in the prairie provinces, the economic advantages of increased size of operation are everywhere clear. The emotional attachment to the family farm, and its political implications, show few signs of weakening, even in a region where the population is now dominantly urban in distribution. However, the most significant political decisions affecting prairie agriculture are likely to be those made by the federal government, and the outlines of those policies which aim to create larger, market-oriented, farms are already clear. On the prairies, further retiring of land economically little suited for agriculture is likely to continue. Greater diversification regionally and increase in farm size will also continue. The continued expansion of the cultivated area, and the massive foreign sales of wheat which characterized much of the 1960s, may well mark the end of an era.

4 Reorganization of the Economy since 1945

BRENTON M. BARR

The economic development of the prairie provinces proceeded initially on the basis of agriculture, and during the first three decades of the twentieth century a significant share of Canadian gross capital formation was composed of prairie farm and transport investment. Since the 1930s, however, growth in production has been chiefly related to mineral extraction, while the need for agricultural labour has been more than halved by improved farming practices. The result has been a radical shift in employment structure, along with equally radical shifts in population distribution and the location of economic activity. Rural-urban migration and rapid urbanization have become major trends; Winnipeg's domination of the urban system has been greatly reduced; the role of the railways has been diminished under the impact of road, air, and pipeline transportation; the plains area has become a significant petroleum producer; and the Canadian Shield has witnessed major increases in the production of non-ferrous metals (Figure 4.1). In short, the region has become an important supplier of fuel and raw materials to Canadian and American manufacturers, and the prairie economy has become polarized around the Edmonton-Calgary corridor and the Winnipeg metropolitan area.

Without economic development based on mineral discoveries, the prairie provinces in 1972 could have been a sparsely settled agricultural region primarily dependent on the cruel vagaries of the international grain market and on the small Canadian market in which prairie agricultural produce faces both domestic and foreign competition. It has been suggested, for example, that if petroleum had not been discovered, the largest Alberta export after World War II might well have been people (Gray 1969). Hanson (Gray 1969) has estimated that without the petroleum and mining industries, the population of the prairie provinces in 1966 would have been only half of their 3.4 million: the petroleum industry alone accounts for more than a million of the additional people. The discovery of oil at Leduc, Alberta, in 1947 symbolizes the beginning of the reorganization of the prairie economy, and no single incident was more

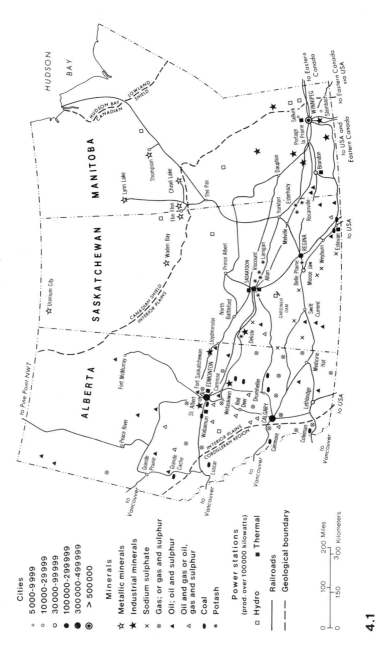

Cities

- ○ 5000–9999
- ○ 10000–29999
- ○ 30000–99999
- ● 100000–299999
- ● 300000–499999
- ◉ >500000

Minerals

- ☆ Metallic minerals
- ★ Industrial minerals
- × Sodium sulphate
- ⊗ Gas; or gas and sulphur
- ▲ Oil; oil and sulphur
- △ Oil and gas or oil, gas and sulphur
- ● Coal
- ＊ Potash

Power stations
(prod. over 100000 kilowatts)

- ☐ Hydro ■ Thermal

—— Railroads
–·–·– Geological boundary

	0	100	200 Miles
	0	150	300 Kilometers

4.1
The Prairie Economic Region

significant. The subsequent discoveries of major supplies of oil and natural gas provided the basis for considerable industrial development within the prairie region. Petroleum reserves in excess of the requirements of future local demand, plus the willingness of the Canadian government to allow exports, stimulated massive inputs of exploration and development capital, with ensuing spillovers into other sectors of the economy. The growth of the petroleum industry and the creation of employment in manufacturing and service industries have allowed the prairie economy to diversify away from the previous agricultural base, and the exploitation of coal, potash, sodium sulphate, sulphur, nickel, and wood pulp resources has helped to broaden the primary industrial base.

Resource developments distinguish the past quarter-century of prairie economic history from preceding periods. Post-war growth in North American consumption of increasingly distant supplies of raw materials has encouraged integration of the prairie economy with the major market areas of Canada and, particularly, of the United States. Economic reorganization, however, has not been a complete panacea for previous uncertainty and instability. Regional prosperity is restrained by surplus wheat crops and competition from other regions in marketing raw materials and attracting new investment capital. Regional development problems are increasingly manifest in the need to apply governmental incentives and assistance to promote industrial investment within the region. Recent events suggest that the self-sustaining development of many prairie industries, based on independent private capital, is being superseded by government financial support for so-called 'hothouse industries' in which, it appears, total benefits are less than the actual costs. Despite the key role played by private investment in the petroleum and non-ferrous metal industries, recent government assistance to the pulp and paper, fertilizer, meat packing, and petrochemical industries suggests that the demand for many prairie products is not growing quickly enough to ensure sufficient provision of new industrial employment.

Population

Changes in the occupational structure of the region have been accompanied by spatial shifts in the distribution of population, particularly to urban centres. Despite an impressive rate of urbanization since 1941, however, the urban proportion of the prairie population is still below the Canadian average of 74 per cent. Approximately three-fifths of the prairie population now lives in urban areas, although in Saskatchewan the proportion is still less than 50 per cent. The rural-urban balance of Manitoba and Alberta approximates that of Canada as a whole, suggest-

Table 4.1   Population growth in metropolitan areas, 1931–71

	1931	1941	1951	1961	1971
POPULATION					
Winnipeg	284,129	299,937	354,069	476,543	534,685
Regina	53,209	58,245	71,319	112,176	138,956
Saskatoon	43,291	43,027	53,268	95,564	125,079
Calgary	83,761	93,021	139,105	279,062	400,154
Edmonton	79,197	97,842	173,075	337,568	490,811
POPULATION GROWTH					
(per cent)	1931–41	1941–51	1951–61	1961–71	
Winnipeg	6	18	34	12	
Regina	9	22	57	24	
Saskatoon	1	24	79	31	
Calgary	6	50	101	43	
Edmonton	18	77	95	45	

Source: Dominion Bureau of Statistics, *Census of Canada*, 1941, 1951, and 1961; 1971 data from advance releases by Statistics Canada.

ing that a large segment of both the prairie and the Canadian population has come to be associated with a more diversified economic base. The distribution of prairie population in urban places reflects the relative importance of small agricultural and resource-extractive towns (under 10,000 population) and the dominant administrative, distributional, and service functions of the five largest regional centres (over 100,000 population). Centres of intermediate size (30,000 – 99,999 population) with a strong secondary manufacturing base are not yet significant in the system of prairie cities. A low density of population in the region encourages major manufacturing enterprises to serve the prairie market from plants in central Canada, the United States, or overseas countries.

Mineral discoveries since 1946 have provided employment for displaced agricultural labour and have offered opportunities for migration of population both within the region and from other regions. Urban population has grown in those places which offered the best access to resource sites and which permitted internal and external economies of scale in commercial and industrial activities. Prior to World War II, the Canadian Census recognized Winnipeg as the only metropolitan area between southern Ontario and the lower mainland of British Columbia. Development of petroleum resources quickly raised Calgary and Edmonton to metropolitan status and the addition of potash had a similar, although less spectacular, impact on Regina and Saskatoon (Table 4.1). Winnipeg has grown at a much slower rate since 1941 than the other large prairie cities,

Table 4.2   Concentration of population in the metropolitan areas, 1931–66

	1931	1941	1951	1961	1966*
PER CENT OF PROVINCIAL POPULATION					
Winnipeg (Manitoba)	40	41	46	52	53
Regina (Saskatchewan)	6	6	8	12	14
Saskatoon (Saskatchewan)	5	5	6	10	12
Calgary (Alberta)	11	12	15	21	22
Edmonton (Alberta)	11	12	18	25	27
PER CENT OF PRAIRIE POPULATION					
The five metropolitan centres	23	24	31	41	44
PER CENT OF PRAIRIE URBAN POPULATION					
The five metropolitan centres	77	80	76	70	70

Source: Dominion Bureau of Statistics, *Census of Canada*, 1961 and 1966.
*Change of urban definition.

although its importance to the concentration of population in Manitoba has increased significantly (Table 4.2).

The increase in population since 1931 has been very unequal in the three prairie provinces. Only in Alberta has the rate of increase approximated the Canadian average. Since World War II, the population of Manitoba has grown at approximately one-half to two-thirds the average national rate, while Saskatchewan has achieved only one-third of the national rate. Intraregional differences in urban growth since 1941 reflect the relatively large urban population base which already existed in Manitoba in 1941, the extremely low urban component in both Saskatchewan and Alberta in 1941, and the differential ability of each province to provide non-farm employment.

Recent mineral developments have encouraged urban growth patterns which add stability to some previously settled agricultural regions. Petroleum, potash, and sodium sulphate extraction have affected urban activities in Saskatchewan within the region previously devoted to farming, as has the petroleum industry in much of southern and central Alberta. In addition, the latter industry has been responsible for considerable growth in small communities in the Rocky Mountain foothills and throughout northwestern Alberta. In the Shield district of Manitoba and Saskatchewan, the extraction of non-ferrous metals has created new aggregated patterns of urban settlement. Similarly, growth of the pulp industry has occurred in all three provinces and has encouraged the growth of urban places in the forested regions beyond the limits of traditional prairie settlement. The expansion of the railway network since 1931 summarizes much of the spatial impact of new resource developments.

The Prairies and Canada

A comparison of the prairie economy with that of Ontario, the most industrialized province of Canada, or with Canada as a whole, facilitates the identification of its key attributes. The prairie dependence on primary industry parallels that of Canada, which has only recently become a moderately advanced industrial power but which continues to depend on international sales of industrial raw materials and agricultural products for a large share of its foreign exchange and export sales. However, internal and external economies of scale, direct access to Canadian and American markets, and proximity to supplies and enterprise in adjacent American industrial regions have permitted manufacturers in south-central Canada (the Quebec-Windsor corridor) to compete successfully with producers in other Canadian regions and to dominate much of the Canadian manufacturing economy. This dominance is based on processes which are increasing in strength and significance with each decade. Having reaped the benefits of an early start, the manufacturers in this corridor successfully limit the possibilities for major expansions of manufacturing in the prairie region. Growth of population depends on the creation of industrial employment and the development of markets. The presence of a lucrative job market in central Canada encourages prospective employees to seek training and employment there, and this, in turn, ensures the continued concentration of the Canadian market. Furthermore, in modern commercial economies the major share of industrial output is consumed by other manufacturing establishments, which further promotes industrial concentration in central Canada. Many factors suggest, therefore, that self-sufficiency is not possible in most Canadian regions except in the manufacture of low-value, bulky, or perishable commodities, commonly found in large urban places and in centres which serve regional markets. It is no surprise to find that manufacturing in the prairie region is dominated by light industry and by custom fabrication related to resource extraction (Figure 4.2).

The relative importance of the tertiary sector in the aggregate prairie economy is identical with that of the Canadian and Ontario economies, but the secondary sector is only half as important as in either Ontario or Canada as a whole (Figure 4.3). During the period 1951–61, while the relative importance of secondary industry was declining in Ontario, the proportion in the prairie economy increased slightly. However, approxi-

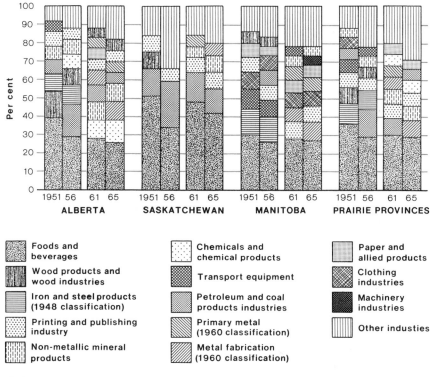

Per cent

1951 56 61 65   1951 56 61 65   1951 56 61 65   1951 56 61 65
ALBERTA   SASKATCHEWAN   MANITOBA   PRAIRIE PROVINCES

Foods and beverages

Wood products and wood industries

Iron and steel products (1948 classification)

Printing and publishing industry

Non-metallic mineral products

Chemicals and chemical products

Transport equipment

Petroleum and coal products industries

Primary metal (1960 classification)

Metal fabrication (1960 classification)

Paper and allied products

Clothing industries

Machinery industries

Other industies

4.2

Value Added by Manufacturing

(Source: D.B.S. "Manufacturing Industries of Canada." Various years.)
Note: a = 1948 Standard Industrial Classification.
      b = 1960 Revised Standard Industrial Classification and New Establishment Concept.

mately 30 per cent of the prairie labour force in 1961 was employed in primary industry, compared with only 10 per cent in Ontario and a Canadian average of 14 per cent.

Employment

The prairie labour force increased between 1951 and 1961 at the same rate as the Canadian average (22 per cent). Alberta with 38 per cent growth, however, far exceeded Saskatchewan (8 per cent) and Manitoba (15 per cent), and even surpassed Ontario (27 per cent). Except for Alberta, the prairie provinces have not shared in the general Canadian industrial expansion which has occurred since 1945.

The relative rates of growth in primary, secondary, and tertiary sectors

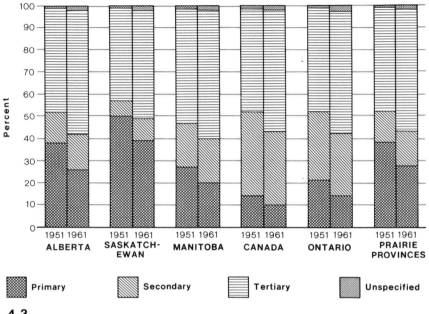

4.3

Distribution of Labour Force

(Source: Census of Canada 1961, Bulletin SL-1, "Labour Force")

suggest that Alberta's economy has grown as quickly as that of Ontario and Canada as a whole. This also suggests that Alberta has become more stable in the post-war period because of the relative decline of primary industry. The change in Alberta has matched or exceeded that of Canada's chief industrial province, suggesting that structural changes in Alberta's economy probably indicate that it is proceeding faster toward economic stability than the other prairie provinces. However, the economic base of Alberta's growth is narrower than that of Ontario since the latter has better access to national and foreign markets for manufactured commodities. Edmonton and Calgary have a similar market disadvantage compared with manufacturers in metropolitan Winnipeg.

Despite Manitoba's more developed secondary sector and wider economic base, the record for recent years shows that the personal income per capita in Manitoba has been consistently below the Canadian average. Although resource-based industries such as petroleum may be potentially unstable in the face of eventual exhaustion of resources, their short-term impact as reflected in personal income is impressive. The anomaly shown

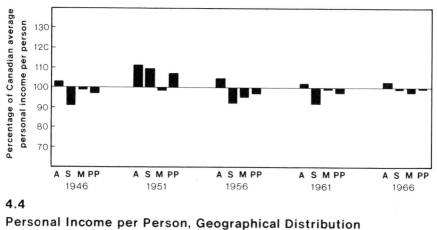

4.4

Personal Income per Person, Geographical Distribution
The figures shown for each year are three-year averages.

(Source: D.B.S. "National Accounts, Income and Expenditure." Various years.)

by the prairie economy is that, on the one hand, stability and long-term growth are sought but, on the other, a district with a more diversified economic base has shown an unsatisfactory income record. When export markets for prairie commodities are strong, personal income per capita is above the Canadian average and the region prospers. However, when international markets soften, personal income becomes markedly inferior to the Canadian average. If the Manitoba economy is to serve as an example of the form of diversification which the other provinces might seek, then the advent of a more diversified economic base throughout the region should create stable levels of personal income which are slightly below the Canadian average (Figure 4.4).

Changes in composition of the prairie labour force suggest that considerable diversification has occurred in recent years. However, because of the uneven substitution of capital for labour among the various sectors of the economy, the magnitude of economic change is not fully reflected in employment statistics. Only in Manitoba does the value added by manufacturing exceed that contributed by primary industry (Figure 4.5). The economy of Saskatchewan continues to rely on agriculture, and that of Alberta depends on mining (petroleum). The nature of goods produced in the prairie region varies considerably from province to province. At the close of World War II, all prairie provinces depended on agriculture for the largest proportion of provincial production, whereas only Saskatchewan now retains its agrarian dominance.

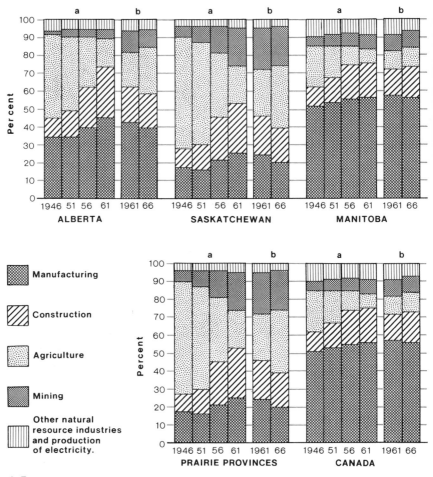

4.5

Census Value Added 1946-66

(Source: D.B.S. "Survey of Production". Various years.)
Note: a = 1948 Standard Industrial Classification.
     b = 1960 Revised Standard Industrial Classification.

Markets

To the extent that heavy dependence on primary industry constitutes potential instability, the prairie economy is still potentially more unstable than that of either Ontario or Canada as a whole. Only Manitoba is close to the average Canadian situation, and its lower dependence on primary industry reduced the economic impact of a recent slump in the international demand for prairie cereal grains. Thus, Operation LIFT (Low Inventories for Tomorrow) of the Canadian government in 1969–70 was of very small consequence to Manitoba farmers, whereas it was of direct importance to those in Saskatchewan and numerous districts of Alberta. Conversely, Saskatchewan potash producers have recently had to impose quotas on themselves, and have brought home the danger of relying heavily on one mineral as a panacea for economic growth. Similarly, American decisions on oil import quotas, which have been pressed for by American producers who seek to exclude cheaper Canadian oil from their domestic markets, have been viewed with considerable apprehension in Alberta. The ease with which oil and natural gas can be transported, the absence of a large regional market on the prairies, and a structure of freight rates which favours movement of raw materials instead of finished products have reduced the impact which these fuels might have had on associated industrial development in the region. Some manufacturers of petrochemicals and other products, such as ammonia, who located near gas fields in central Alberta in the early 1950s before the export pipelines were constructed, have found that the cost of distributing finished products to eastern markets has increased faster than the cost of assembling natural gas at these markets. For those prairie manufacturers whose major markets are located in central Canada and abroad, a raw-material-oriented location can be a serious disadvantage. Furthermore, the increasing shortages in the supply of natural gas in the United States will probably increase the price which Canadians pay in the prairies as well, thereby further decreasing the attraction of an apparently cheap fuel. With its worsening strategic position and diminishing domestic supplies of petroleum, the United States would probably not curtail the import of Canadian petroleum. However, the possibility of substituting other fuels for oil and gas, or other sources of supply for prairie petroleum, is of critical importance to the Saskatchewan and Alberta economies. In 1968, for example, the value of petroleum production in these provinces was one billion dollars. Recent development and exploration outside the region suggest that the boom of post-war exploration might be over, and that investment in leases and infrastructure may be far less important in the next decade.

Is the economic base of any region completely free from market uncertainty and unemployment? Diversity of the economic base and flexibility in the response of manufacturers to changes in market demand are major stabilizing factors. However, a large proportion of the prairie manufacturing economy is oriented to regional markets and is thus dependent on wealth generated by primary industry. Manufacturers of fertilizer, for example, have recently seen markets decline because of adverse sales conditions for grains. Transportation costs tend to restrict the market area of this industry to the prairie provinces and adjacent American states. But the manufacturers were beguiled into overdevelopment by successful wheat sales to the USSR and China in the early 1960s, which led to predictions of ever-increasing export markets for prairie wheat. The analysts failed to anticipate that poor harvests in the USSR and China would be followed by intensive government schemes to improve domestic agricultural production in those countries. With approximately two-fifths of the world's wheat surplus, Canadian farmers are not heavy consumers of fertilizer and many plants are operating far below productive capacity. The normally low rate of fertilizer use was further decreased in 1970 by a one-year programme of the federal government to lower wheat inventories by promoting summer-fallow and seeding to grass. A partial extension of this plan for three years is intended to encourage forage production and the growth of the prairie livestock industry. Prairie fertilizer manufacturers depend on international sales of grain and on a poorly planned, uncoordinated, and rather conservative prairie agricultural environment. Similarly, in many prairie districts, sales volumes of wholesalers and retailers depend directly on the ability of the agricultural sector to market grain and livestock. Such businesses, which suffered in 1969 and 1970, benefited from a strong international demand in the 1970–1 crop year for Canadian oilseeds and feed grains, and from the announcement by the federal government of measures to encourage the stabilization of income from grain sales and so reduce the impact of fluctuations in world market prices.

Transportation costs are also very important to steel-pipe producers because of the relatively low value of the commodity in relation to its weight and bulk. Pipe mills are located in Edmonton, Calgary, Camrose, and Regina, but full production depends upon continuous expansion of the prairie transmission networks and a steady flow of orders for pipe. The last discovery of a major oilfield in Alberta occurred in 1966, and the pipe manufacturers expect that a large proportion of their production in the 1970s will be used in the construction of a giant oil and gas transmission line from the Arctic via the Mackenzie Valley to American markets. Pipe manufacturers are subject to many forces from outside the prairie provinces, such as the demand for petroleum and gas, foreign petroleum prices, the discovery of new petroleum deposits, the export policies of federal and provincial governments, and the actions of integrated steel companies located outside the region. For example, a long delay in the Canadian Energy Board's approval of new exports of natural gas caused severe disruptions in pipe production for at least one major manufacturer in 1970. In addition, with only one prairie integrated steel-pipe operation, most pipe manufacturers purchase steel outside the region from producers who also manufacture pipe in direct competition with prairie mills.

Primary and Semiprocessed Commodities

Diversification of economic activity has weakened the strength of agriculture, which served as the basis of regional identification for the first half of the twentieth century. The prairies continue as Canada's most important agricultural region, but the region itself is no longer strictly agricultural. The three prairie provinces rely heavily on agriculture and resource-extraction for the generation of wealth. These activities pose common problems to all three provinces. In addition to cereal grains, potash, crude oil, and natural gas, the prairie region is an exporter of sulphur, nickel, copper, uranium, zinc, sodium sulphate, wood pulp, and coal. Most of the final manufacturing and consumption of these commodities is carried out elsewhere. Whether agricultural or industrial, the basic commodities produced in the prairie region are exported in unprocessed form and are important for economic activity in other regions. Prairie oil and gas sales, for example, in the past fifteen years appear to have helped to foster basic industrial development in central Canada and many districts of the United States at the expense of Alberta and Saskatchewan. Recent approval by the Canadian government of extensive gas sales to the United States is additional evidence that a basic function of the region will continue to be the supply of raw materials to domestic and foreign industrial regions.

Prairie industrial growth has relied heavily on capital from other regions, particularly from the United States. However, the presence of foreign ownership and control of resources has not led to the localization of wealth in prairie urban places or to the establishment of banks, investment houses, and other financial institutions with a commitment to foster continuous development in the region. Prairie resource developments have not created general economic growth akin to that of the southwestern United States or California. Rather, their situation is more like that of the north-central states of the American Great Plains. The distribution of population in Montana, North and South Dakota, Wyoming, Nebraska, and northern Minnesota is sparse and discontinuous. Their chief economic advantage appears to be their inclusion in the large domestic American market although, like their Canadian neighbours, they suffer a distance handicap in reaching national and foreign markets. Unfortunately for both areas, other national regions are better endowed with resources, underwent earlier resource development, and provided more opportunities for the establishment of financial institutions and a strong population base. Despite its strong financial community earlier in this century, the economic and financial leadership of Winnipeg has been steadily eroded by Toronto. The central business districts of all prairie cities except Calgary attest to the absence of a strong financial or business community resembling that of Toronto or Montreal. American interests, which have dominated the Canadian oil and gas industry, have concentrated their Canadian corporate offices in Calgary, partly because of the initial development of the petroleum industry in southern Alberta, and partly because of the easy access to related companies in the western United States.

The prairie region is in a state of flux arising largely from uncertainty about the strength of the continuing demand for its raw materials. Alberta has benefited from the sale of oil and gas leases and from royalties derived from petroleum production, but the revenues now accruing to the government of Alberta from oil and gas leases are approximately one-quarter those obtained during the late 1960s. Production of conventional petroleum in Alberta is expected to peak within several years. Petroleum discoveries in other regions, such as the north slope of Alaska and the Mackenzie delta of the Northwest Territories, plus the anticipated discovery of major quantities of petroleum in the Gulf of St Lawrence, constitute a threat to the resumption of intense oil exploration in the prairie provinces. Predictions of continued prosperity from petroleum sales usually include the development of the Athabasca tar sands. These major deposits of bitumen in north-central Alberta are estimated to contain over 300 billion barrels of recoverable synthetic crude oil. Although recovery processes are technically feasible, the loss of over 30 million dollars by Great Canadian Oil Sands during the first three years of operation in Fort McMurray suggests that development by other companies is still premature. And yet, the need to secure sources of supply by integrated oil companies seems to augur well for additional development of the tar sands before the end of the twentieth century, and the establishment of new settlements in parts of Canada's mid-north. The tar sands will be expensive to develop, however, and will have to compete internationally with other projects for investment capital.

The fate of resource developments in the prairie provinces depends on the profitability of ventures in other world regions. In the case of nickel, for example, Canadian technology and Canadian funds are helping to develop mines in New Caledonia, Guatemala, the Dominican Republic, the Philippines, Australia, and Indonesia. Anticipated increases in nickel production throughout the non-communist world could lower Canada's share from 65 per cent in 1970 to approximately 38 per cent in 1980. In 1954, Sherritt Gordon Mines Ltd., of Toronto, using a unique hydrometallurgical process, began to refine nickel in Fort Saskatchewan, Alberta. As other firms located in the area, a small chemical complex developed. Subsequently, Sherritt Gordon has licensed refineries in the Philippines and Australia to recover nickel and cobalt from lateritic ores. The possibility of overseas applications of a technology which was originally developed for refining ore in the prairie region appears to reduce the likelihood of rapid expansion of nickel processing and associated employment in the region.

The inability of other world regions to supply sufficient materials can be of great benefit to the prairies. Following the dieselization of the railways in the early 1950s, for example, the prime demand for prairie coal was in thermal electric plants. Now, however, high-grade coking and steam coal deposits in the Alberta foothills are being mined for the Japanese market. Experiments are also being carried out to determine the feasibility of shipping coking coal to central Canada to fill gaps left by shortages of coal from traditional American sources. However, the demand for coking coal in other regions such as Japan could be affected by the availability of alternative cheaper supplies in Australia, South Africa, or, upon the conclusion of joint Soviet-Japanese agreements, Siberia.

International market conditions also affect other prairie commodities. The demand for potash, for instance, has been far below expectations. Canadian potash production is concentrated in Saskatchewan, where the nine producers are prorationed to operate at approximately half their rated capacity. A considerable market for Canadian zinc is found in the United States, but the planned removal of toxic elements from gasoline could reduce prairie sales of zinc to American markets by perhaps one-third.

Recognizing that many prairie commodities are sold in unstable markets, the federal and provincial governments have attempted to expand the primary industrial base by providing financial assistance in the form of loan guarantees, regional development incentives, and assistance with resource inventories. This assistance must be seriously questioned, however, because it has been carried out on a piecemeal basis and not as part of a long-range development plan in which alternative forms of investment were carefully evaluated. The pulp and paper industry, in particular, has recently been artificially stimulated. Wood pulp is manufactured in Hinton (Alberta) and Prince Albert (Saskatchewan). Newsprint is manufactured at Pine Falls (Manitoba). These mills commenced operation in 1965, 1967, and 1927 respectively. The Hinton pulp mill has access to high-grade cellulose fibre and has enjoyed considerable assistance with inventories and harvesting areas from the provincial government. During a recent decline in the market for pulp, the American parent company closed similar mills in the United States but continued operations in Hinton to maintain an amicable relationship with the provincial government. The newsprint mill at Pine Falls does not work at full capacity because of a chronic shortage of spruce pulpwood. The Saskatchewan government has a 30 per cent interest in the Prince Albert mill, and planned to participate heavily in the construction of another mill at Meadow Lake, 120 miles north of Prince Albert, by providing 30 per cent of the equity capital and by guaranteeing loans of 70 million dollars to the pulp company. The Canadian government also promised a 12 million dollar incentive grant, but unfavourable economic conditions led the Saskatchewan government to cancel plans for this mill in 1971. In the same vein, the Manitoba government is reported to have lent 90 million dollars to four companies which are constructing a forest complex, including a pulp and paper mill, at The Pas. The Canadian government is expected to contribute a further 15 million dollars in area development grants. Yet many representatives of the Canadian business and financial community have questioned various transactions associated with this project. The Manitoba government appears to have participated unwisely with developers whose background and motives are not fully understood. It is a flagrant example of the desperation with which some provincial governments attempt to encourage regional development. And, as a final case, plans were announced in December 1970 for a pulp mill near Grande Prairie, Alberta. This project will cost approximately 80 million dollars but will receive nearly 12 million dollars as development incentives from the Canadian government. In addition, the Alberta government will provide over 3 million dollars for pollution control. Approximately 400 people will be employed directly in the plant and 300 in woodland operations. The cost to government of helping to create 700 jobs is in excess of $21,000 per worker.

In view of the cost of creating each job and the apparent subsidization of production costs through government loans and guarantees, should not some sectors of the prairie economy remain undeveloped and the investment capital be allocated to more profitable locations in Canada until such time as market demand can ensure the profitability of each project?

Prospects

The prairie provinces are particularly subject to extraregional economic and political forces. They appear to be an appendage to the food-deficit nations of the world and to the major industrial countries with reserves of hard currency. Provincial and federal governments are desperately attempting to create jobs through subsidies on resource extraction and the encouragement of primary manufacturing. Transportation costs continue to handicap attempts to set many prairie commodities in distant national and world markets. Other regions and countries compete for development capital and population, thereby casting constant uncertainty over continued prosperity and the evolution of a strong regional market on the prairies. Furthermore, prairie commodities are subject to market influences which are largely beyond the control of the Canadian government. The world price for coking coal, for example, can be countered only through subsidized production or transportation costs. The prices of too many prairie commodities are determined by world market conditions for long-term price stability to be possible. Prairie residents often accuse central Canadian manufacturers of hiding behind high tariff walls in order to capture the domestic Canadian market. However, manufacturing oriented to the domestic market is relatively easy to protect whereas export-oriented production is extremely vulnerable to market fluctuations.

What are the prospects for the prairie provinces in the 1970s? Intraregional disparity of population and industrial growth will probably increase. The Systems Research Group of Toronto has predicted that, by the year 2000, Ontario, British Columbia, and Alberta will increase their share of the national population from 51 per cent to 61 per cent. Manitoba, Saskatchewan, and the Atlantic provinces are expected to show little growth, and thus to decline in their relative positions. Net outmigration of population from Manitoba and Saskatchewan appears to be a reality as the natural increase of population continues to exceed the creation of new jobs. As internal and foreign migrants settle in central Canada, many economies of production will result in the growth and diversification of manufacturing in that region and will further enhance its competitive position and stability. Government incentive programmes for the development of primary and secondary manufacturing will probably affect all parts of the prairie region and, indeed, many other districts of Canada as well. Industrial growth with or without government assistance will continue to rely heavily on imports of foreign capital, particularly from the United States. Increasing government participation and regulation appears as a likely aid, although not a panacea, to problems of overproduction and unsold stocks of many prairie commodities. Manufacturing and distribution activities will further concentrate in major urban centres and, in some cases, will further centralize their operations to one city. In 1970, for example, federal forestry research services in the prairie region were centralized in Edmonton and most activity in other cities was curtailed. Gulf Oil is now completing a large refinery in Edmonton which will render their plants in Calgary, Moose Jaw, Saskatoon, and Brandon obsolete. In addition, many Canadian military forces at small prairie bases are being redeployed to major bases such as Namao (Edmonton). Small service centres, dependent upon resource extraction, will continue to dot the landscape, particularly in the mid-north, but production of most prairie commodities will remain dependent on extraregional demand. The region is similar to many others in that domestic governments cannot guarantee stability in the price of export commodities. Prairie developments related to potash, coal, tar sand, wood pulp, and agricultural processing, however, suggest that the federal and provincial governments will play an increasingly strong role through a variety of financial incentives in assisting individual producers to increase their ability to offer primary commodities at competitive prices in world markets.

The major function of the prairie economy – to supply raw materials to manufacturing regions – will not change until natural population increase provides sufficient markets to support major secondary manufacturing establishments in the region, or until regional resources become depleted.

5 The Population

THOMAS R. WEIR

The prairie provinces are vast and sparsely settled. They contain one-sixth of Canada's population, but their total of 3.6 million in 1971 was still small in relation to their area. Even in the oecumene, which comprises roughly the southern third of the provinces (Figure 5.1), the average density is only 6 persons per square kilometre. The rural density is very much lower still, and there is no escaping the fact that the region is strikingly underpopulated. As has been made plain through the preceding chapters, the small numbers and their uneven distribution are a product of the interplay of environmental limitations, an interior location, recent settlement, and a resource-oriented economy. The population character-istics, in their turn, set major constraints on the type and scale of economic development that is possible, though for many people who have moved to the prairie provinces their emptiness is one of their greatest attractions.

The pattern of population distribution, both within and beyond the oecumene, is closely tied to the physical base of the provinces. In the latter regions, apart from a scattering of small, isolated communities of native people, the few settlements are completely a response to resource exploitation. Within the oecumene, there is a close relationship between the distribution patterns of population and agriculture, which in turn reflects the soil and climatic patterns described in Chapter 1. Figure 5.1 shows how the urban places of all sizes become more widely spaced as rural density declines, a reflection of the rural service role which they largely fill. The recent economic shifts are revealed too, though, parti-cularly in the close spacing of settlements in the Edmonton-Calgary corridor, and the associated ridge of rural population density. The linear form was determined historically by the alignment of railways, and coincides with the black soil, mixed farming zone, but the density gradient has been steepened sharply by the polarization of the new economic activities of the past 20 years. Winnipeg has had a similar effect on the population distribution pattern of southern Manitoba. The new economic order and the urbanization trend are becoming securely imprinted on the

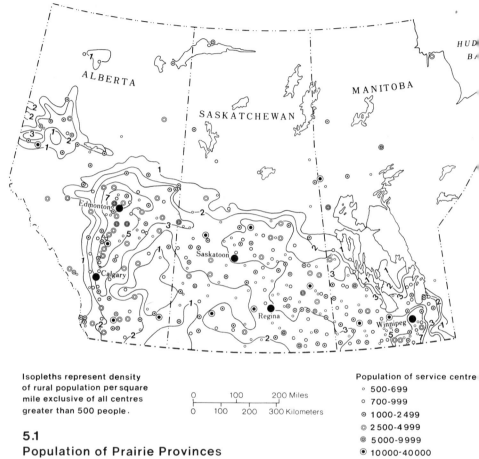

Isopleths represent density
of rural population per square
mile exclusive of all centres
greater than 500 people.

0   100   200 Miles
0   100   200   300 Kilometers

5.1
Population of Prairie Provinces
(Source: Census of Canada 1966)

Population of service centre
° 500-699
○ 700-999
◎ 1000-2499
◉ 2500-4999
◉ 5000-9999
◉ 10000-40000
● Over 40000

population map, and the simplicity of the old relationships is beginning
to be supplanted.

RURAL-URBAN RELATIONSHIPS

According to the Census of Canada, 'urban' refers to a city, town, or
village having more than 1000 population, including its urbanized
fringes. However, this definition does not take into account a large
number of hamlets or small towns which may have as few as 50 inhabi-
tants. The census includes such places under 'rural non-farm' population.
And yet, even hamlets of 50 persons usually have a post office, a general

Table 5.1 Urban and rural population, Canada and prairie provinces, 1966 (in thousands)

	Total population	Rural farm	Rural non-farm	Total rural	Total urban
Canada	20,014	1913	3375	5288	14,726
Prairie Provinces	3,381	718	541	1259	2,122
Manitoba	963	160	157	317	646
Saskatchewan	955	280	207	487	468
Alberta	1,464	278	178	456	1,008

Source: Dominion Bureau of Statistics, *Census of Canada*, 1966.

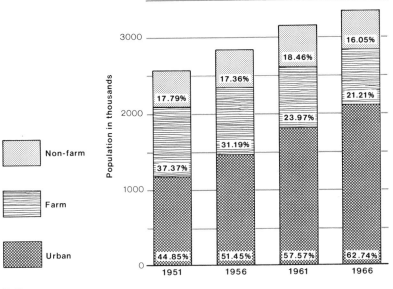

5.2

Rural-Urban Population Change 1951-1966

Expressed as a percentage of the total population.

(Source: Census of Canada 1966)

store, a garage, a grain elevator, and, by reason of a beverage concession, a hotel of sorts. Although they are but tiny prototypes of urban places, they do perform the functions of central places and should properly be considered 'urban.' For lack of other data, however, the census definitions must be accepted (Table 5.1).

Figure 5.2 indicates that between 1951 and 1966 the total population of the prairie provinces increased at a fairly uniform rate from 2,547,770

Table 5.2   Percentage change in population, 1951–66

	Total	Urban	Rural non-farm	Rural farm
Manitoba	+24	+47	+28	−25
Saskatchewan	+14	+85	+14	−30
Alberta	+56	+123	+12	−18
Prairie Provinces	+32	+85	+14	−24

Source: Dominion Bureau of Statistics, *Census of Canada*, 1951 and 1966.

to 3,381,613, a gain of 24 per cent. However, the urban component grew most rapidly, and increased its share from 45 per cent of the total in 1951 to 62 per cent in 1966. At the same time, the farm sector declined from 37 per cent to only 21 per cent, and the rural non-farm proportion (chiefly hamlets and towns having less than 1000 people) remained nearly static. Within the rural non-farm sector, the larger settlements grew in number (from 187 to 252) and in total population (from 254,098 to 402,889) but the smaller centres, while remaining the same in number, declined in population from 160,793 to 156,466. This marks the trend toward the gradual disappearance of the small service centres which cannot compete with the greater variety of services provided by the larger centres, assuming that distance is not a critical factor. Improvements in the highway network have had a particularly striking impact on accessibility patterns, and have hastened the trend to centralization in the higher-order urban places.

Even more affected by declining population is the farm component indicated in Figure 5.2. Although the reasons are primarily economic, the decline has far-reaching social implications which will be discussed later.

In addition to the shifting importance of the urban, farm, and non-farm components of the paririe population, there have also been pronounced differences in the rates of change among the three provinces (Table 5.2). Manitoba had a moderate increase of 24 per cent between 1951 and 1966, owing largely to the slow growth of Winnipeg, which constitutes half the total population of the province. The rate of farm depopulation (25 per cent) was intermediate between that of Alberta and Saskatchewan. Alberta's substantial growth rate (56 per cent) was chiefly the great leap forward of Edmonton and Calgary. Small towns showed a slow rate of increase (12 per cent) in contrast to Manitoba's (28 per cent). Saskatchewan, the most rural of the provinces, witnessed a slow rate of increase (14 per cent), which was explained by the high

Table 5.3   Urbanization in the prairie provinces, 1951–66

Census period	I	II	III	IV	V	VI	VII
To 1951	–	15	11	–	3	26	37
1951–6	4	19	13	0	3	30	43
1956–61	2	20	11	1	4	37	48
1961–6	2	21	9	1	5	44	53

I   Number of towns reaching 5000 population in each period.
II   Total number of small cities of 5000 to 100,000.
III   Small-city population as a percentage of total prairie population.
IV   Number of small cities reaching 100,000 population in each period.
V   Total number of large cities of 100,000 and over.
VI   Large-city population as a percentage of total prairie population.
VII   Urban centres exceeding 5000 as a percentage of prairie population.

rate of farm depopulation. Although its urban growth rate was high (85 per cent), its urban base is relatively small and by 1966 it was still only 49 per cent of the total population. By contrast, in both Manitoba and Alberta two-thirds of the population is urban.

From Table 5.3 it is possible to deduce urbanization trends from 1951 to the present. Small cities (5000 to 100,000) increased from 15 in 1951 to 21 in 1966, though their proportion of the total population remained about the same. This in large part was due to the moving of Regina and Saskatoon out of the 'small city' category. As a result, although large cities increased by only two in number, their total population increased very rapidly, from 37 per cent of the prairie population in 1951 to 53 per cent in 1966. In effect, most of the growth recorded in the prairie region took place in Edmonton, Calgary, Saskatoon, Regina, and Winnipeg. It is ironical that Winnipeg, the one-time metropolis of the prairies, should be the slowest growing of Canada's big cities, while Calgary and Edmonton have had the highest rates of increase. The petrochemical developments in the western part of the prairie provinces touched off a chain reaction of industrial activity and urban growth, and Winnipeg, at the apex of a rail network, lost many of the advantages that formerly adhered to its gateway location. As the significance of prime location was lost, so Winnipeg's growth rate was slowed, a trend which now seems to be affecting Edmonton and Calgary as well, as the peak of the oil boom has passed (Table 4.1).

ETHNIC ORIGINS

Although 90 per cent of the population of the prairie provinces is Canadian-born, the ethnic origins of the people are both numerous and varied.

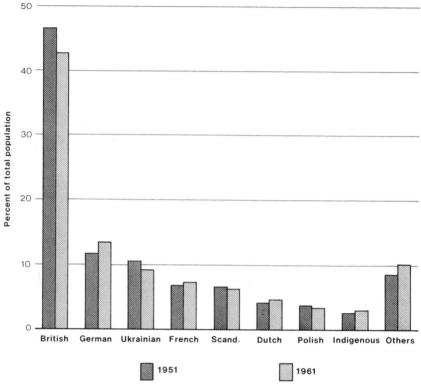

5.3

Proportional Ethnic Composition 1951 and 1961
(Source: Census of Canada 1961)

More than half, excluding those of British descent, originate from some part of Europe (Figure 5.3). People of British stock are the dominant group (43 per cent), their numbers decreasing in the order English, Scottish, and Irish, which is comparable to Canada as a whole. They are disseminated fairly uniformly throughout the region, with heavier concentrations in southwest Manitoba and central Alberta. The remaining groups are not sharply differentiated in numbers, although people of German origin are most numerous (14 per cent). This, in part, is due to a large representation of Mennonites and Hutterites, distinct cultural groups which are included with Germans for census purposes. Immigrants *directly* from Germany have tended to disperse widely and be assimilated quickly into the predominantly Anglo-Saxon culture of western Canada. Their tendency to increase is a reflection of sustained immigration after

World War II. On the other hand, the main period of Ukrainian and east European immigration was at the time of initial prairie settlement, from 1896 to 1913. There was another influx in the 1920s but there has been virtually no immigration to add to their numbers since 1940. As at the time of initial settlement, the present distribution of people of Ukrainian descent favours the northern margin of the oecumene, from Edmonton to Manitoba. They continue to have a strong attachment to the land, and many have persisted in farming under difficult environmental circumstances. The French group is the next largest (7 per cent), but it is much smaller than in Canada as a whole (30 per cent). It includes descendants of the Métis, of mixed French and Indian parentage, as well as descendants of group colonization from Quebec. Although people of French origin are widely dispersed, concentrations occur in the Peace River district, to the east and west of Edmonton, along the fringe of rural settlement in Saskatchewan, and in parts of Manitoba, especially along the Red River. Collectively, Scandinavians form the fifth group. They include Danes, Swedes, and Norwegians, but the earliest to settle in any numbers were Icelanders who took up land on the southwest shore of Lake Winnipeg in 1875. From here they spread across the west, usually as farmers. Although the smallest of the Scandinavian groups, the Icelanders have retained the strongest ethnic identity.

Of the group referred to as 'indigenous,' all are people of Indian origin except about 150 Eskimos at Churchill. The Indians, as distinct from the Métis, have largely chosen to remain on reserves and participate in the rights of their ancestors established by treaty. The Métis, although related to the Indians, are not registered by the Department of Indian Affairs and have no treaty rights. There are an estimated 113,500 of them in the prairie provinces and they follow a way of life not unlike that of the Indians. The Treaty Indians numbered 91,246 in 1967 and were distributed among 159 bands. About two-fifths of the bands live within the prairie oecumene, gaining a livelihood by farming (especially cattle raising) and seasonal labour. Those beyond the oecumene have adapted to the environment of the Canadian Shield, living chiefly from trapping, fishing, hunting, and gathering, with some labour in mines or on construction sites. Improved standards of health and a high birth rate resulted in a 91 per cent increase in the Indian population between 1947 and 1967, with a consequent increase in the social and economic problems which beset the Indian people. With a standard of values quite different from that of the material culture by which they are surrounded, but in which they have little part, the Indians find shelter on the reserves as wards of the government. They have had little incentive to change their pattern of

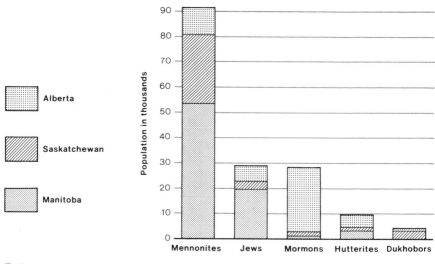

5.4
Religious Minorities 1961
(Source: Census of Canada 1961)

living, and their contacts with the white man's ways have usually contributed to degeneracy, especially in urban centres. The problem of relating the Indians to twentieth-century living continues to defy solution. They remain an ethnic anomaly in a culture with which they have almost nothing in common and into which they do not choose to assimilate.

The problems of Indian identity are set in sharp relief by a number of even smaller minority groups who have managed to preserve their own distinct value systems (see Figure 5.4). By a tenacious adherence to intense religious convictions, these groups have maintained a clear social identity and, in some instances, have resisted assimilation at any level.

The Mennonites are the most numerous and the oldest of the group colonists. They were German-speaking people who migrated to Manitoba from southern Russia in 1874. They established two settlement nuclei from which daughter colonies migrated westward to Saskatchewan and Alberta. Initially they settled in compact villages, using the land in common, but each family held title to a quarter-section granted under the Dominion Lands Act. In time, disaffection for a close village life set in and the original villages gradually disintegrated. Out-migration occurred, mostly to Mexico and Paraguay, for those who wished to retain their agrarian base and their religious and cultural traditions. Those who remained have largely become integrated with Canadian cultural practices, and may be recognized only through religious affiliation. It is estimated, for instance, that 80 per cent of the Mennonites in Manitoba are urbanized. The old emphases on village life and German language, which were originally seen as unifying forces, have largely disappeared.

By contrast, the Hutterites, who came to Manitoba in 1918 from South Dakota, have retained their cultural identity, their religion and language, and a strict communal way of life. A rapidly growing group with a high birth rate, a low death rate, and no tendency toward assimilation, their numbers have increased from 625 to 12,137 (1969). This would be considered minor indeed were it not for their land acquisition policy and their practice of forming new sister colonies when the parent colony reaches 125 to 150 persons. In 1969 there were 130 colonies: 45 in Manitoba, 18 in Saskatchewan, and 67 in Alberta. Strictly communal, the individual owns no real property, is subject to an exact routine, and is governed by a set of rules designed to shield him from the unwholesome influences of the society about him. Able to operate as a corporation, and able to benefit by laws in a society which gives them equal rights with other farmers, Hutterites are able to compete very well as an agricultural colony. They encounter varying forms of resentment as a result, particularly since they make no contribution to community life. For example, although they must comply with provincial education laws, Hutterite youth grow up with a minimum knowledge of Canadian institutions. Ill-prepared to take his place in a competitive society, the Hutterite rarely deserts the colony. To meet the rapid increase in numbers of Hutterites, all three provincial governments have now imposed certain restrictions on the purchase of land. Nevertheless, public suspicion and some resentment prevail toward the Hutterites, who are considered to be a seventeenth-century anachronism.

A third minority group, the Mormons, have retained a strong areal concentration even without the unifying advantage of a distinctive language. The Mormon settlement of southern Alberta was begun in 1886 and, like the Mennonites and Hutterites, this group made a unique impress on the rural landscape. Their settlement forms and agrarian life have been greatly modified, but as a group they still predominate in their areas of initial colonization.

DYNAMICS OF CHANGE

In the 15-year period 1951 to 1966, the population of the prairie provinces increased by 24 per cent. This rate varied regionally as well as among rural, small urban, and large urban settlement (Table 5.2), in

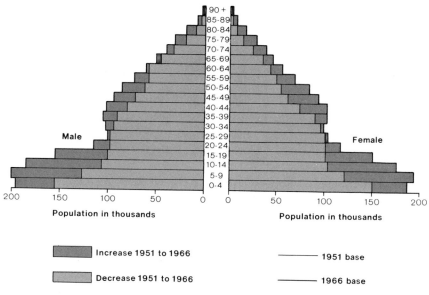

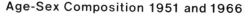

5.5

Age-Sex Composition 1951 and 1966

(Source: Census of Canada 1951 and 1966)

accordance with the demographic and social forces which establish the dynamics of change.

Historic Trends

Figure 5.5 consists of superimposed population pyramids for 1951 and 1966. It therefore measures the population change in each five-year age group over the 15-year period. Only two do not show increases. The decrease in 65 to 69 year old males perhaps reflects a falling-off of births during the 1890s because of an agricultural depression and an exodus of young men. However, it should be noted that for all age groups from 45 years up to 80 years there was a surplus of males. This corresponds to the large influx of settlers into the prairie region between 1880 and 1914, as land was taken up under the generous provisions of the Land Acts. The large proportion of single men in the population of that period caused the abnormal bulge accompanying the period of initial settlement.

The more obvious constriction in both pyramids occurred after World War I and continued through the depression of the 1930s. War casualties and the influenza epidemic affected the subsequent birth and death rates.

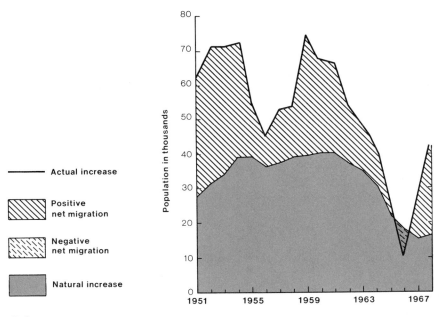

5.6
Demographic Trends in the Prairie Provinces
(Source: D.B.S., Vital Statistics, Catalogue No. 84-202, 1951 to1968)

The age group 25–29 calculated from the 1965 base was born in the late thirties when the birth rate reached an all-time low. This is also indicated by the decrease in the female population of that age group between 1951 and 1965. The very substantial increase in the age groups from 5 to 20 years old was brought about by the 'baby boom' following World War II. The 1951 population was low because of the effect of depression and war on births, but the surge after 1951 is most strikingly obvious. The beginning of the trend to delayed marriages and smaller families is apparent in the 0–4 years group. The 1951 base is substantial, but 15 years later the dropping off of the infant population is quite apparent. This trend has accelerated since.

Recent Trends
The relationships of the three demographic factors (births, deaths, and migration) for 1951 to 1960 are shown in Table 5.4. Over the prairie provinces as a whole, the crude birth rate rose from 26 per 1000 in 1951 to 29 in 1954 and then entered on a slow decline until 1964. Thereafter

Table 5.4 Population change through natural increase and net migration by five-year intervals, 1951 to 1966

	Total increase and rate of increase*			Natural increase			Net migration		
	1951–56	1956–61	1961–66	1951–56	1956–61	1961–66	1951–56	1956–61	1961–66
Manitoba	73,499 (10)	71,646 (8)	41,380 (4)	73,684	76,006	70,340	−185	−4,360	−28,960
Saskatchewan	48,937 (6)	44,516 (5)	30,163 (3)	86,030	86,294	75,691	−37,093	−41,778	−45,528
Alberta	183,615 (20)	208,828 (18)	131,259 (10)	120,961	144,234	134,607	+62,654	+64,594	−3,348
Prairie Provinces	306,051	324,990	202,802	280,675	306,534	280,638	+25,376	+18,456	−77,836

Source: *Canada Year Book*, 1969, p. 156.
*As a percentage, in parentheses.

it fell steadily to 19 by 1968, after which it began to level off. At the same
time the death rate remained constant at about 8 per 1000. The effect
on natural increase is clearly evident from Figure 5.6. Growth by natural
increase was very high through the 1950s and early 1960s, with a peak
rate of over 2 per cent per annum. Thereafter it diminished rapidly, so
that in 1968 natural increase contributed only 20,000 to the regional
increase as compared with 40,000 in the peak year of 1961. The actual
annual increases have been even more erratic, with sharp crests and
troughs which reflect the effects of migration, and particularly the great
variability in foreign immigration. Overriding the short-term fluctuations,
however, both Table 5.4 and Figure 5.6 provide evidence of a general
downward trend, and an actual net out-migration during the 1960s.

Also from Table 5.4 it is apparent that the demographic controls operated quite differently in each of the prairie provinces. Population growth
in Saskatchewan and Manitoba was at a much smaller rate than in Alberta
during the period 1951–66. However, the latter province has also experienced a much reduced growth rate and a net out-migration flow during the
sixties. Through the 1950s there was little regional variation in the effect
of natural increase: it was high in all the provinces. The variation in rates
of provincial increase, then, was a direct response to the migration trends:
the strong in-flow to Alberta, the strong out-flow from Saskatchewan, and
Manitoba's near balance. Since 1961, Saskatchewan has experienced a
sharper decline in natural increase than the other two provinces, but the
differential has been partially balanced by the first large out-migration in
Manitoba and the disappearance of Alberta's positive flow.

The net migration involved both Canadian-born and foreign-born residents of the three provinces. In Manitoba, Canadian-born constituted the
out-migrants and foreign-born the in-migrants. Only in Manitoba was
there a near balance of these two elements in the decade 1951–61. Saskatchewan received very few foreign-born migrants, whereas Alberta
received both Canadian-born and foreign-born migrants, the latter being
three times greater than the former (George 1970). Manitoba gained
most of its Canadian in-migrants from Saskatchewan, followed by the
Atlantic provinces; Saskatchewan showed a small in-flow from Prince
Edward Island; and Alberta drew from every province except British
Columbia, but especially from Saskatchewan and Manitoba. In terms of
out-migration, Manitoba lost most heavily to British Columbia, and then
to Alberta and Ontario; Saskatchewan lost heavily to Alberta and British
Columbia, and nearly all migration from Alberta was into British Columbia. From this pattern it is obvious that the two factors guiding the
destination of out-migrants were geographic proximity and affluence.

Moreover, of the total interprovincial migrants between 1956 and 1961, the proportions going to urban destinations were as follows: Manitoba 80 per cent, Saskatchewan 70 per cent, and Alberta 80 per cent.

Rural-Urban Migration

Any discussion of the dynamics of change within the prairie oecumene would be incomplete without mention of the strong rural to urban flow. This, combined with the influx of immigrants from Europe and the United States, largely explains urban growth into the sixties. To an extent this growth has been at the expense of rural farm areas (Figure 5.2). Between 1951 and 1966 farm population losses were 24 per cent compared to urban gains of 85 per cent. It has been estimated that in 1961 about half the non-agricultural labour force over 25 years of age in Saskatchewan was raised on farms (Abramson 1968). The intraprovincial movement from farm to town or city has been the strongest, and most far-reaching migratory movement in Canada since the thirties. Of 7237 townships in the prairie provinces, 79 per cent lost population during the period 1941 to 1956. Those which gained usually had an urban centre within their boundaries. In Manitoba, of 1071 townships comprising the oecumene only 18 per cent gained population over the same period, whereas the remainder suffered losses ranging from 10 per cent to well over 40 per cent. Losses of farm population have been largest in Saskatchewan, followed by Manitoba and Alberta in that order. The only rural areas of the prairie provinces to gain population lay on the outer fringe of the oecumene, especially in the Peace River district, and in the irrigated districts of southern Alberta. Otherwise, only in the towns and cities was growth recorded. The continued drain of people from the farming areas goes on with little tendency to diminish and will continue as long as an economic imbalance exists between town and country. Government rehabilitation projects and the expenditure of welfare money in backward rural areas may retard the movement away from farms, but such measures are not enough to produce a significant abatement.

Large and widespread losses among farming people are directly traceable to two causes: reductions in the number of farm families through migration, and decreases in the size of farm families due to declining birth rates and the migration of sons and daughters at an early age. In Manitoba's Interlake district, between 1941 and 1961, farm population declined 39 per cent; the number of families decreased by 30 per cent and the average family size by 15 per cent. The latter was in large part due to absolute decreases of 46 per cent in the number of young children and of 64 per cent in the number of young adults. The loss of nearly two-thirds of the potential homemakers augurs poorly for the future population of this farming area. These rates of change apply only to a sample area in Manitoba, and are not typical of all farming areas in the prairie provinces, but they may be taken as indicative of a widespread rural trend.

Causes and Consequences of Rural Depopulation

The basic cause of rural depopulation is economic. There are contributory causes of a social nature, but most have an economic base. Two sets of forces producing migration are involved. Negative forces arising from economic insecurity serve to 'push' the farm operator from the land. These are complemented by positive forces which 'pull' the discouraged farmer or his children to alternative employment which promises both an easier and a higher standard of living. Associated with these are such amenities as educational advantages for the children, more attractive living conditions for the family and especially for the wife, and better health and community services. Notwithstanding, the city way of life is regarded with suspicion, and easy living has to be weighed against the advantages of life on the farm. The decision to move is in the end very complex.

It is generally conceded that movement of people represents the search for greater economic security and is a normal adjustment of population to the resource potential of a given area. The Census of 1961 indicates that 17 per cent of farms in the prairie provinces were non-commercial; i.e., were incapable of supplying a reasonable income, either because of the physical quality of the land or because they were too small. A study of 65 townships in Manitoba's Interlake district showed that the average farm size increased from 105 hectares to 160 hectares in the period 1941 to 1961. In the process, 59 per cent of the quarter-section farms disappeared and 10 per cent of those with two quarter-sections. Associated with this was a corresponding increase in the number of larger farms. The increase in farm size and simultaneous elimination of the small farm is a response to the need for more land to utilize modern farm machinery to advantage. Capital investment is high, and only large units are capable of meeting expenses. The farmer who lacks managerial skill, who regards farming as a way of life rather than a business, is sure to be eliminated in time. Many small operators are unable to obtain credit to expand, or find it impossible to secure land at a price they can afford. The marginal farmer is unable to withstand reverses due to climate, insect pests, or animal diseases, and finds himself incapable of recovering financially from a 'poor year.' In a study of rural migrants in Saskatoon in 1966 it was found that 41 per cent desired a higher standard of living for their families,

31 per cent sought better educational opportunities for their children, and 23 per cent wanted less anxiety and strain (Abramson 1968). Isolation from medical centres (especially for those of advancing years), secondary schools, and advanced training facilities, and opportunity for social contacts were some of the reasons given for moving.

Rural depopulation sets up a chain reaction. As families move, the municipality's tax base is reduced and services are curtailed. Local schools are closed and long trips on the school bus to consolidated schools in large towns then follow. The school and church were formerly centres of social activity, and with their disappearance a sense of community is lost. The rural landscape is scarred by abandoned farmsteads, schools, and churches. Isolation causes many to move to nearby hamlets when the type of farming permits. In many instances the local service centre finds itself incapable of competing with the larger centre at a distance. Mobility is greater and the economy-minded farmer travels farther for his services and his social contacts. Life becomes more impersonal. Many who found employment in small towns are forced to move in turn to bigger centres. Hamlets are eliminated, buildings abandoned. Life on the isolated farmstead, where the houses are two or more kilometres apart, becomes lonely and unattractive. Disintegration of rural communities removes many of the satisfactions of rural life which formerly helped to compensate for a lower standard of living.

The Canadian government has taken cognizance of such problems stemming from rural migration and has taken steps to make rural living more attractive. Large sums have been made available for rural rehabilitation through educational and training programmes, the development of local resources, retiring marginal lands from cultivation and opening community pastures, extending credit and management advice, and providing schools, roads, hospitals, and other services. Although such programmes will retard the trend to urban centres, the flow from the land to the city will continue until something approaching a balance in economic opportunity is achieved. Meantime, young people, better trained than their parents, will continue to seek the advantages they perceive in an urban society.

6 Changing Forms and Patterns in the Cities

P.J. SMITH

The historic rural base of the prairie provinces is now very remote to most of the people who live there. Prairie society has been transformed by the new urban world, particularly in the five largest cities which now contain 50 per cent of the provinces' population and have themselves been transformed by their increased size and prominence. Much of the change can be attributed simply to growth. The combined populations of the five cities in 1951 was about 800,000: twenty years later it was 1,700,000. Yet growth did not just mean more of what there was before. It was a time of changing attitudes towards urban living and urban building, and a time when city planning became an accepted medium for change in the urban environment. Even more than size, these new concepts and new ways of living have put light-years between the prairie city of the 1970s and its progenitor of the generation before.

THE PATTERN OF PHYSICAL EXPANSION

The simplest change is the expansion of the urbanized area. Individual cities have doubled, trebled, even quadrupled in size in twenty years or less, pushing out their boundaries in surging advances. In a very general way, the cities have come to be girdled by modern forms of development, the recent enclosing the older, but a simple concentric model would be hopelessly inadequate as an explanation of patterns of expansion. None of the cities has grown by a steady process of annular accretion (Figure 6.1). The typical pattern is for new building to be channelled into development corridors, along which it advances rapidly. When further progress along these corridors is barred by a major constraint, in all probability the focus of attention will shift to another development area in another part of the city, forcing back an earlier barrier which now seems less of an obstacle. Even in a period of rapid growth, some sectors of the city may remain virtually frozen.

There are many factors which influence the shape of the prairie cities

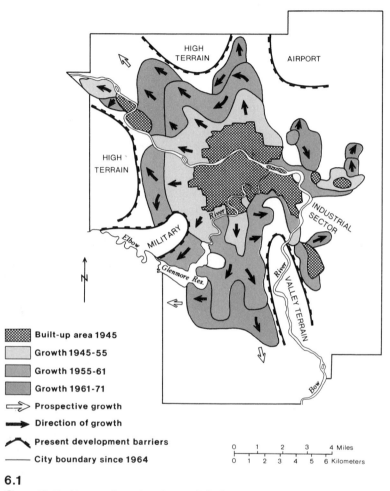

6.1
Growth Patterns for the City of Calgary 1945-1971

and their directions of growth at any given time. At least two, though, have been major controls throughout. The first is the dominant role of superior residential areas in the total residential pattern, and the way in which they have moved along the rivers in distinct narrow sectors, pre-empting the prime building sites. In the prairie environment, the river valleys provide almost the only strong physical interest – variety of relief, attractive vistas, handsome stands of trees, and, sometimes, access to water. The effect is most immediately striking in Winnipeg, whose rough cruciform outline is still built on spines of high-quality housing along the

Red and Assiniboine rivers. The Elbow River in Calgary, the North Saskatchewan in Edmonton, and the South Saskatchewan in Saskatoon all have the same attraction. Even in Regina, the most featureless of the cities, the modest channel of Wascana Creek has something of the same impact.

The second outstanding control is exerted through the public utilities, particularly water and sewer systems, streets and railways. The physical and financial feasibility of opening new areas for development has been in the forefront of public land policy since 1946, and with good reason. The great settlement wave of the first decade of the century crested extravagantly in the cities. In the land boom years of 1908 to 1912, tens of thousands of hectares were subdivided around each of them, sufficient for populations five or ten times larger than those of the day. Much of it remained paper subdivision, with a title registered to an absentee owner in Toronto or Chicago or London, but much was also serviced with graded streets, sewer and water systems, electricity, and even streetcar tracks. The local authorities went deeply into debt in the process, but the collapse of the land boom meant that the services lay idle through vacant, tax-delinquent land. The payment for the period of madness was long and hard. As late as 1960 it was not uncommon to find established residential districts with no paved streets, no sidewalks, and no street lighting. It is no wonder, then, that the new boom days of the fifties and sixties have seen a more cautious approach to public spending. This has imposed an order and compactness on the form of urban expansion that did not exist in earlier periods.

If there is an exception, it is Winnipeg. Because of its earlier rise to prominence and the early fixing of municipal loyalties, Winnipeg has always had a more fragmented structure and a more dispersed pattern of growth. Both characteristics have persisted to the present. By 1921, when the other cities had populations of 25,000 to 60,000, Greater Winnipeg already had a quarter of a million people and the central city was largely encircled by independent communities. These formed foci for subsequent growth and change, and so have persisted as substantial nuclei.

The other prairie cities, by contrast, have historically been mono-nuclear. Edmonton and Calgary each had neighbouring towns but they remained small and were overwhelmed in the growth of the central cities. The old town centres have persisted as business nuclei but, unlike in Winnipeg, they are completely overshadowed by the new nodes which have emerged during the great physical expansion of the past twenty years. For Edmonton and Calgary the mid-fifties was a crucial time. They were in the first freshet of major growth, new housing was being built at inconvenient distances from their central areas, and the notion of the planned shopping centre was just coming into common currency. There was also public initiative in reserving municipal land as shopping centre sites in new residential areas. Private initiative was quick to follow, and the past fifteen years has brought a continuing evolution of a finely structured hierarchy of service centres. Regina and Saskatoon show an early stage of the same trend, as do the outlying parts of Winnipeg.

Despite the appearance of new business nuclei in the four western cities, the central business districts have continued to grow and become more specialized as the unchallenged apexes of their particular central place networks. Only in Winnipeg, the oldest and most slowly growing city, has there been serious concern about the decline of the central area. But here, too, the trend seems to have been checked by recent massive investment.

Other service facilities which were historically situated in or near the city centre have likewise been confronted with the choice of expansion on a relict site or relocation. Edmonton shows various solutions to this dilemma. For instance, one central hospital has been rebuilt on its old site, and a second has removed completely to a suburban location; the University of Alberta, on a site too small for its present population, has absorbed an adjoining residential district; and the municipal airport was destined for closure ten years ago, a move that was foiled indefinitely because the new international airport was built too far out of the city. At the same time, new services have almost invariably gone to peripheral locations. The new universities in Calgary and Regina, regional parks and golf courses, and a variety of institutions have sought sites where space and good road access have been substituted for centrality.

Industry and related businesses have shown the same pattern of change even more clearly. Relocation from traditional central situations is now a well-established trend, particularly for wholesaling firms but to a considerable degree for manufacturing firms as well. Problems of outmoded plants, inadequate sites, and poor road access have been compounded by the physical spread of population and the dispersal of business centres. A central situation does not necessarily provide the greatest convenience any more, particularly with the almost universal dependence on motor traffic for movement within the city. New or incoming businessess are largely limited to peripheral sites, though this has not produced a radical change in the industrial location pattern. There has always been a strong spatial association between the railway network and industry in the prairie cities, and this has persisted despite the modern emphasis on truck transport. The physical expansion of industrial areas, then, has been accomplished by a combination of infilling and outward growth along the railway tracks.

In other ways, the rise of the automobile has had a most striking impact on the form of the prairie city. The whole pattern of physical growth and dispersal would have been impossible without it. More specifically, the new street forms have become a notable element of the urban landscape. Each of the cities has at least the rudiments of a ring road system with intermeshing radials focusing on the central area. It is very difficult to impose this system on the underlying grid matrix, but in the outer districts it is usually possible to provide a clear arrangement of service, feeder, and through streets. An ordered hiararchy of urban streets is therefore emerging, slowly and expensively.

THE PATTERN OF POLITICAL EXPANSION

As an inevitable consequence of physical expansion, the political boundaries of the cities were quickly overrun. Questions of the political and administrative organization of urban space have therefore been as critical, in their way, as the questions of physical organization, and the two have constantly interacted. There have also been a variety of responses to the problems created by overspill from the central city, though all have had the common goal of trying to unify the total urban area for purposes of physical planning and development control.

Even before the great period of physical expansion, the political space pattern of the prairie cities was fragmented. Long before the ground within the central cities was fully occupied, building had occurred in adjacent municipalities, some of which were themselves legally constituted as villages or towns. In a few instances, the non-central communities were created for a special purpose, as in the railway settlement of Transcona to the east of Winnipeg, or in Bowness which was promoted as a garden city at the end of an umbilical tramway line from Calgary. But, for the most part, the overspill was simply a low-density scatter of detached houses, mostly of inferior quality, taking up a small part of the land which had been subdivided in the land boom. By World War II, then, the cities were loosely encircled by partially urbanized municipalities which were inadequately equipped to provide a decent standard of urban services or to cope with the demands of a period of rapid urban growth. Winnipeg provided the only exceptions to this general pattern, in the exclusive residential town of Tuxedo on the Assiniboine River, laid out ca. 1910 to a design of the Olmsted brothers, and in the City of St Boniface which faces the City of Winnipeg across the Red River and achieved a notable economic and social stature in its special role as a French-speaking enclave.

The earliest response to the problems of controlling urban expansion in divided jurisdictions came as a direct result of the resumption of growth in the early post-war years. It took the form of metropolitan planning commissions in Winnipeg (1948), Edmonton (1950), and Calgary (1951). The central cities were empowered to join with their fringing municipalities for the purpose of preparing an over-all master plan and so ensuring that all new development should be subject to uniform control.

The early years of metropolitan planning were not particularly successful, largely because the commissions lacked effective authority. The Alberta reaction was to legislate new powers for the planning commissions and to refine the relationship between them and the city administrations in a two-tier planning structure. Thus, the planning commissions embrace very extensive areas around the two principal cities, and they are completely responsible for detailed planning within these areas but outside the city boundaries. They are also responsible for establishing the policy goals and guidelines for the development of the whole region, thus setting a framework within which the city planning departments flesh out the details for the two city areas. This organization has not always worked smoothly, but it has helped to produce one of the chief successes of planning in Alberta, the containment of Edmonton and Calgary and the virtual elimination of urban sprawl. Indeed, Edmonton and Calgary are almost European in the abruptness with which they march with the rural landscape.

In Winnipeg, by contrast, there was only a single level of planning from the beginning, and the metropolitan planning commission was only one of a series of ad hoc bodies which was vainly trying to persuade the nineteen urbanized municipalities to adopt a coordinated approach to land development. The Winnipeg solution was the creation, in 1960, of a new tier of local government, the Metropolitan Corporation of Greater Winnipeg (Figure 6.2). This agency was to have sole responsibility for development control and the provision of essential services to the whole urban area. It was a major step, but unfortunately it still left ample scope for conflicts of interest between 'Metro' and the individual municipalities, and these bedevilled it throughout the decade of its life. In 1971, the Manitoba government took the ultimate step of amalgamating all the metropolitan municipalities into a single unit.

The notion of single, unified administrations is by no means new in the prairie provinces. Except for Winnipeg, all the cities have regularly extended their boundaries to take in raw land in advance of development.

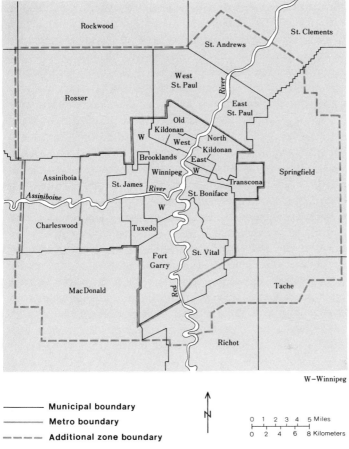

W—Winnipeg

——————— Municipal boundary

——————— Metro boundary

— — — — Additional zone boundary

0 1 2 3 4 5 Miles

0 2 4 6 8 Kilometers

6.2

Municipal Boundaries in the Winnipeg Area, 1964

In the Saskatchewan cities, it has long been the pattern for the city limits to contain the entire urban area and for no development to be permitted on land which is not incorporated within the city. The boundaries are simply extended as new building land is needed. This does away with the necessity of a second tier of planning or administrative organization, but it also reflects the isolation and small size of these cities.

The situation in the larger Alberta cities has not been that simple. Though never as hemmed in as Winnipeg, both Edmonton and Calgary have had their extensions constrained at times by the existence of independent towns and strongly urbanized rural municipalities. In Calgary's case, these difficulties were removed in a series of boundary extensions between 1961 and 1964. The whole urban area became incorporated with the City of Calgary, and a huge fringing tract of raw land was included as well (Figure 6.1). The city planning department thus became the sole agent for urban planning and development control in the metropolitan area, and the possibility of conflict of interest with the regional planning commission was virtually eliminated. The latter became primarily concerned with preventing urban development beyond the city boundaries and with overseeing the extra-urban space needs of Calgary residents, such as regional parks and summer cottage sites.

In Edmonton, on the other hand, the regional planning commission has remained much more directly involved with urban planning. In the first place, the City of Edmonton has not been allowed to extend as comprehensively as the City of Calgary. All the contiguous residential areas have been annexed, but the city has not been allowed to take in some very large-scale industries on its eastern margin. These yield so great a tax return to their county that several attempts at annexation have been rejected. In the second place, Edmonton is the only metropolitan area in the prairie region where a deliberate policy of satellite town development has been followed (Figure 6.3). The notion first surfaced about 1950 and was a clear reflection of the British experience of the planners who were active in the Edmonton area at the time. It quickly became established policy that a number of small towns in the vicinity of Edmonton should be allowed to grow, and so restrain the physical expansion of Edmonton city. Tied in with this was the notion of partial green belts to prevent the closest satellites (St Albert and Sherwood Park) from being overrun. The Commission had no power to direct development to the satellite towns, so its main concern has been to create a permissive environment through its master plans and its development approval procedures. The main investment interest has been in commuter housing, particularly as serviced land has fallen into short supply in the city and land prices have risen steeply. For the most part, then, the satellite towns are dormitories for Edmonton. Only one, Fort Saskatchewan, has a substantial industrial base.

THE ORGANIZATION OF NEW SPACE – RESIDENTIAL

In the first phase of urban development, the standard form of subdivision was the grid, or some simple modification of the grid, adjusted to the rural subdivision base. This consisted of long, narrow river lots in Winnipeg and part of Edmonton, and mile-square sections elsewhere. In

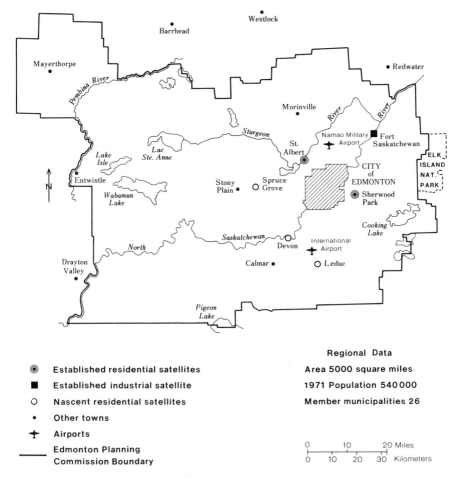

Regional Data

Area 5000 square miles

1971 Population 540 000

Member municipalities 26

◉ Established residential satellites

■ Established industrial satellite

○ Nascent residential satellites

• Other towns

✈ Airports

── Edmonton Planning
    Commission Boundary

6.3
Edmonton in Its Planning Region

Alberta and Saskatchewan a rigid, completely rectangular pattern resulted, and even Winnipeg's greater variation was a reflection of the
irregularity of the underlying matrix, rather than a departure in principle.

Very little of the land which was subdivided and serviced in the land
boom years was actually built on then. It provided a reservoir of building
sites for over forty years, and up to sixty years in some parts of Winnipeg,
though the pattern of development through the 1920s and 1930s tended
to be badly fragmented. By the late forties, when the demand for house
lots was again on the upswing, there was still much vacant land in the

tattered fringe zone, and the early post-war construction simply continued the long process of in-filling, until the old supply of serviced land was at last exhausted. In parts of Winnipeg the process is still going on, but in other cities it has already been replaced completely by a very different pattern of residential organization. Although it is a crude generalization, the new forms represent a third band of development around the prairie cities. The first is small and fast disappearing: it consists of a dense development of very similar, usually two-storey houses built ca. 1910. It gives way to a much broader band of housing, very mixed in style, age, and quality because covering a development span of 40 to 50 years. Then, in the expanding outermost band there is a reversion to the consistency of the cities' first age. The houses obviously belong to a different period but the areas have the same air of having been built as a unit in a short time. Now, however, the rectilinear street pattern of the two inner zones is supplanted by the curvilinear patterns which are the distinctive mark of the modern subdivision.

The new style of subdivision made its first appearance in Edmonton and can be attributed to one man, Noël Dant, the city's first professional planner, who took up his appointment in 1950. With Dant, the notion of the neighbourhood unit became known in the west, and it soon became the established development model in Edmonton and Calgary. It also became the model, though less distinctly, in Regina and Saskatoon, but in Winnipeg it appears only in a very debased form. The difference appears to be attributable to the scale of development in the Saskatchewan cities and local government fragmentation in the case of Winnipeg. Much of the residential development in Regina and Saskatoon has simply been a matter of rounding-out existing areas to acceptable boundaries, such as ring roads or railways or city boundaries. As yet there have been limited opportunities for building complete neighbourhoods on virgin sites. In Winnipeg, the scale of development has also been less than in either Edmonton or Calgary, and this smaller total has been dispersed among sixteen municipalities in the metropolitan area. Again, then, the opportunities for building complete new neighbourhoods have been comparatively rare.

One condition which favoured the acceptance of neighbourhood units was the high proportion of municipal land ownership. The only positive outcome of the land boom was that much of the vacant, subdivided land passed to the municipalities in lieu of tax payments. For many years it was a liability, but in the new boom days of the 1950s it gave city councils the power of absolute control over planning and development. Wherever possible, the old subdivisions were cancelled and replaced by the completely new neighbourhood forms, designed on the principle that residential areas should be self-contained in so far as local services are concerned. The neighbourhood, then, is not just a unit of residential organization. It also sets the distribution pattern of convenience shopping, education, and recreation, and structures the highway network (Figure 6.4).

Another innovation with the neighbourhood unit was the notion of mixed dwelling types. The traditional residential building in the prairie cities has been the single-family detached house, and before 1950 almost nothing else was built. It still predominates in new residential areas, but the old pattern of uniformity is slowly giving way, particularly in Edmonton and Calgary. Mixing of dwelling types has conventionally been seen as the means of securing social heterogeneity, an impossible goal which has led to the most serious criticisms of the neighbourhood unit concept. In the Alberta cities, however, there has been no such altruism. Mixing has been used purely as a physical planning device for the protection of investments in single-family properties. The real distinction is not one of social class or even of building form: rather it is one of tenure. Single-family houses are for owner-occupancy and the aim of the planners is to ensure that they can never be devalued through land use or traffic conflicts. Other dwelling types are primarily for renter-occupancy, and the renter has been seen as a transient creature, not deserving the same consideration as the owner-occupier. The least attractive parts of the neighbourhood units have therefore been allocated to rental accommodation. By building duplexes (semidetached houses) or walk-up apartments on the sites fronting the arterial streets or adjoining the shopping centre, the single-family owner-occupiers are given a high degree of visual and functional protection. The fact that these strips are referred to as 'buffer zones' is an explicit indication of the planners' intent.

The early residential mix was limited in conception, function, and building type. Recently, however, these attitudes have moderated, partly as planners and developers have become more aware and more confident of the consumer appeal of diversity, and partly because of changing attitudes among the consumers themselves. Rental accommodation is now in much greater demand than ever before, and less stigma is attached to it. The reasons are by no means completely clear. Economic factors, such as high land prices, building costs, and interest rates, have made homeowning impossible for many people, but there seem to be even more enduring changes in life style which give renting a permanent appeal for the first time in the history of the prairie cities.

The first intimations of this trend emerged in the early sixties, with the construction of occasional small blocks of luxury apartments on the

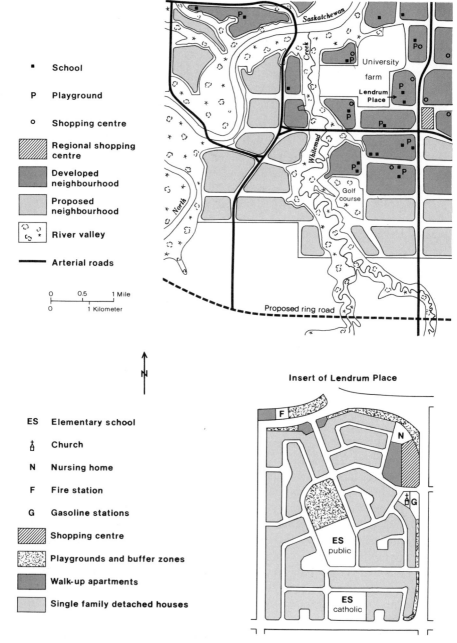

Legend:

- ■ School
- P Playground
- o Shopping centre

Regional shopping centre

Developed neighbourhood

Proposed neighbourhood

River valley

Arterial roads

```
0        0.5       1 Mile
0            1 Kilometer
```

N

- ES Elementary school
- ☦ Church
- N Nursing home
- F Fire station
- G Gasoline stations

Shopping centre

Playgrounds and buffer zones

Walk-up apartments

Single family detached houses

Map labels: Saskatchewan, Creek, University farm, Lendrum Place, Whitemud, North, Golf course, Proposed ring road, Insert of Lendrum Place, F, N, ES public, ES catholic, G

6.4
The Neighbourhood Pattern in Southwest Edmonton
(Modified from Atlas of Alberta)

edges of some of the most expensive neighbourhoods. It was clear evidence that even these bulwarks of privilege could be breached in the interest of the affluent minority who wanted suburban living without home ownership. The trend has not gone unopposed and indeed has sometimes been frustrated by local home-owners' groups. Such rearguard actions show that many prairie urbanites still treasure the goal of home ownership and believe that other dwelling types will depreciate the amenity and value of their properties. Where the neighbourhoods are mixed from the beginning, however, there is less scope for conflict, and this is the contemporary pattern.

The mixed neighbourhood ca. 1970 is very different from that of even five years before. Single-family houses take up less of the area, and are more intricately and subtly mingled with the other housing types. Moreover, these other types are much more varied in style and arrangement. Duplexes and row houses, often built for owner-occupancy as well as for renting, are mixed with various apartment forms, from walk-ups to twenty-storey towers. The principles of cluster housing are widely employed and special attention is being paid to landscaping. The total effect shows a greatly heightened awareness of urban design, an awareness that has never before been part of the prairie ethos.

THE ORGANIZATION OF NEW SPACE – NON-RESIDENTIAL

Although residential use has taken by far the lion's share of new urban space, other very substantial space needs have also been generated. They, too, have been subject to changing attitudes towards development standards and planning. These will be reviewed under the headings of nodal and non-nodal space.

Nodal Space

The chief contributor to nodal organization is the shopping centre, which first appeared in the early fifties as part of neighbourhood unit planning. It quickly fell into a standard pattern of supermarket with a small attached cluster of convenience shops and service outlets. Its form also became standardized – a linear arrangement which is essentially the traditional shopping-street with the addition of ample parking. Its nodal role shows up in the tendency to locate other neighbourhood services (e.g., churches, gasoline stations) and rental housing on adjacent sites.

The process of subcentralization comes through more strongly in the higher-order shopping centres which have been developed outside the neighbourhood pattern. The first generation of district centres was built in the late fifties and contained a department store and forty or fifty other outlets. They were designed for a larger and more specialized market than the neighbourhood centres, but their appearance differed only in scale. The designers had not yet moved beyond the traditional street-front concept.

Since then, two significant things have happened.

(a) There has been a transformation in the design of shopping centres. About 1965 the covered mall appeared as a second generation of district centre. Not only did this improve customer comfort, through climate control and a higher standard of design, it also made for a more convenient and economical layout, with shops arranged down each side of the mall. Almost immediately the older centres were converted to the mall style, with increased floor space and incomparably improved conditions for the shopper. More recently still, the mall design has been used in neighbourhood centres and in the much larger regional centres which are the latest addition to the commercial hierarchy. These have two department stores and up to 75 other outlets and services, many of them highly specialized.

(b) The high-order district and regional centres are now acting as major magnets in the urban matrix. Movement patterns are being completely transformed and the urban highway networks are having to be restructured as a result. The drawing power of these new nodes is being further reinforced by the complementary activities which are now clustering strongly around them. Automobile services of all kinds, cinemas, rival supermarkets, even small shopping centres are being drawn, moth-like, to the flame of the concentrated market. Public and professional services, such as branch libraries and medical clinics, are similarly held. Apartment developers have also been quick to see the advantages of these consumer service nodes, and their locations can frequently be pinpointed by the high-rises which now dot the suburban landscape.

Lesser examples of nodal aggregation are provided by some institutional and transport facilities. The University of Calgary, for instance, was founded on a tract of rolling grassland beyond the limits of suburban housing: in ten years it has spawned an extensive complex of shopping centres, sports arena, major hospital, and dwellings of all types. Airports, too, have had an aggregating role in all five cities: industry, warehousing, and traveller services (e.g., hotels) are the most obvious attracted uses.

Non-nodal Space

Non-nodal activities neither cluster at points of peak accessibility nor lead to an agglomeration of complementary functions. Nonetheless, they produce some very distinctive uses of space, such as industrial districts, highway commercial ribbons, and metropolitan parks.

The notion of planned industrial land development came more slowly and less enthusiastically than that of planned residential development. Partly this reflects a very optimistic approach to industrial zoning: within the broad peripheral tracts that have been zoned for industry, there is an abundance of subdivided land and the in-filling process tends to be very slow. There is also a reluctance to set standards which might deter industry from settling in the prairie cities, and the needs of large space users, in particular, are very difficult to fit into any predetermined mould. To a large extent, then, industrial land development has been fragmented and unorganized. At the same time, though, the principles of industrial estate planning have gained some acceptance in all cities, and most particularly in Calgary where the City Council took the initiative in laying out municipal land for use by industry and distributive trades. Other developers, both public and private, have since followed suit, in Calgary and elsewhere. Although the planned industrial estate in the full sense is not really part of the prairie scene yet, there has at least been a more orderly approach to industrial land subdivision and a higher standard of building design.

If anything, the highway commercial ribbon has created an even greater problem of development control than has industrial expansion, partly perhaps because it is more visible but also because it has accommodated such a variety of uses. Little imagination has been used in the layout of the ribbons, which are simply zones of mixed business use flanking the main access highways. They are disorderly, unsightly, obstructive, and hazardous to traffic, and a perpetual source of nuisance and frustration to the planners. Fortunately, in recent years they have shown little growth other than in-filling, and in Edmonton and Calgary they have been almost completely contained. Strict zoning has helped, but positive location alternatives were also needed. A shift back to the city centre for hotel construction, the rise of the shopping centre on or near the access routes, and the greater use of planned industrial areas – these are now absorbing the functions which once sought sites in the highway business zones. The highway ribbon appears to be as obsolete as the shopping street of an earlier generation.

Metropolitan parks are strongly related to site conditions, and particularly to river valleys. The most elaborate system is being developed along the North Saskatchewan and its tributary ravines through Edmonton, but the same pattern can be observed in Calgary and Regina. Winnipeg's recreation is less oriented to its rivers and more to the large lakes and forest reserves within a 50-mile radius, a pattern which is shared to varying degrees by the other cities. To complement these resource-based parks within and beyond the urban areas, a close mesh of playgrounds has also developed. These are integrated with the residential and school distribution patterns and form an essential element in neighbourhood unit planning.

THE REORGANIZATION AND REUSE OF OCCUPIED SPACE

In an areal sense, the recent changes in the prairie cities have been expressed most largely in peripheral accretion. The most dramatically visible changes, however, have occurred in their central areas, particularly in Edmonton and Calgary. In the other cities the pattern of redevelopment is the same but the process is at a much earlier stage. In Regina and Saskatoon this seems to be a function of size; in Winnipeg it is a result of slow growth.

Redevelopment in the Central Area

All the prairie cities entered the post-war era with strongly established business cores of a generally compact form. Their revived growth was therefore accommodated by horizontal expansion which was permitted over extensive areas by zoning law. In the sixties this trend was replaced by vertical expansion and the in-filling of the extended business zones. The individual buildings and the density of buildings are both rising higher and higher, in a return to compactness on an enlarged scale. The original business nuclei, in the meantime, have survived with little change apart from spot redevelopment and occasional refurbishing. The streets of low, close-packed, narrow-fronted buildings are sadly overshadowed by the glass and aluminium towers, and the signs of neglect and economic decline are evident all too frequently. So far, however, redevelopment has been hindered by the small properties and the difficulty of assembling sufficiently large sites at a reasonable cost for modern building projects.

The continuing vitality of the business cores is in part a result of public policy. In all five cities, a strong business centre is accepted as a basic planning goal, and a great deal of money is being spent to make it work. Much of this is direct investment in land assembly and redevelopment, either for new municipal facilities (e.g., civic offices, public service buildings, libraries) or for eventual reuse by private interests. There are also heavy expenditures which are designed to improve the accessibility and amenity of the central area – the upgrading of access streets, the construction of bridges and freeways, the creation of downtown parks, and (in Calgary) the conversion of part of the main shopping street into a pedestrian mall – all designed to maintain the competitiveness of the central area by enhancing its convenience and appearance. Another closely related policy is that of maintaining a large local population. High-density residential zoning is therefore being used to facilitate apartment redevelopment, not just on the margins of the central area but actually penetrating to the very edge of the retail core.

One aspect of central area planning which has not been successful is civic design, and particularly the integration of individual buildings into a total pattern. The existing planning machinery provides only the most mechanistic powers for controlling building heights, densities, and setbacks, and there is boundless latitude for the developer whose sole concern is to extract the maximum amount of floor space from the available money. The office zone in Calgary shows the result: a dense packing of buildings of varied heights and styles, in which the merits of individual structures are lost in the prevailing disorder. Equally sad is the office or hotel which soars vainly above a wasteland of parking-lots, obsolete buildings, and the leftover oddments of an earlier existence.

On the positive side, municipal and provincial governments are now setting a good example in some of their own redevelopment projects. Where scale, building arrangement, open space, and other design concepts are handled intelligently and sensitively, as in Winnipeg's Centennial and Civic centres, a new level of amenity can be introduced into the city centre. Fortunately, too, some private developers are sharing this concern for urban environment and have been responsible for several well-conceived building complexes. The notion of a comprehensive design for the central area, though, is still an impossible dream.

Reshaping the Commercial Ribbons

While the central areas have flourished and generated ever more traffic, there has been increasing pressure for business development along the access routes. As in the central area itself, the first response was horizontal, but this seems to have been checked, partly by zoning and partly by the alternative of shopping centre location. Again paralleling the central area, the recent trend has been towards more intensive use, in-filling the commercial frontages through the removal of relict houses, and now redeveloping some of the older commercial property. The ribbons are still pre-eminently the home of space-hungry types of retailing (e.g., furniture and appliance stores and automobile sales outlets) but a significant new feature is becoming apparent. As the central areas become more intensively built up, they no longer have an economic place for the small office block. For the present, these are being built in the commercial ribbons, but there is an obvious danger if the trend continues: more intensive use is not compatible with traffic flow, and may eventually force the traffic to be relocated.

Residential Areas in Transition

The third and largest portion of the prairie city to be undergoing massive reorganization is the inner residential area. The trend to residential conversion and redevelopment is now well advanced in Edmonton and Calgary, and it will not be many years before the pre-oil boom cities have almost completely disappeared. A few protected enclaves of high-cost houses will probably survive, but the rest is fast yielding to an apparently insatiable demand for rental accommodation in central locations. In the other cities the same trend is just beginning to surface, but it must be assumed that it will have the same overwhelming effect.

The pattern of residential redevelopment is deceptively simple. There is an obvious relationship with distance from the central area, and this is enshrined in zoning practice. The high-rise towers close to the business district give way to walk-up apartments, first in compact rows and then strung along the access routes. Yet the central area is by no means surrounded in this manner. Redevelopment is channelled into well-defined corridors, seeking out areas of established amenity and good housing quality. River-banks and crests of valley bluffs have high attractiveness, as do tree-lined streets and middle-class districts with well-maintained houses. Residential deterioration does not act as a spur to redevelopment. On the contrary, it is a deterrent, and each of the cities has its zone of residential discard, its inferior central housing which is being avoided in the private redevelopment process.

The spatial pattern is further complicated by the effect of other activity nodes. The growth of the University of Alberta, for example, is forcing the almost complete redevelopment of the old housing areas on Edmonton's south side. The old business nodes have also become foci for residential redevelopment. And increasingly, to the untutored observer, the redevelopment pattern is becoming indistinguishable from the development pattern, as it reaches out along the arterial streets and begins to march with the early neighbourhoods. Apartment redevelopment thus has to be seen as part of a broader pattern of rental housing, and can by no means be isolated from events in the outer reaches of the city. Some important but as yet unanswerable questions follow. Will redevelopment march unabated into the post-war neighbourhoods which were designed for the preservation of single-family housing? Will the new trend to suburban apartments check the pace of redevelopment, or will the suburban trend be stimulated if resistance stiffens in the inner city?

Questions such as these make it plain that the forces of change are by no means spent in the prairie cities. The old forms and patterns may already be doomed, but this does not mean that their replacements will be any more durable. In many ways, the concepts of ten years ago are no less outmoded, though the counter-forces of tradition and privilege are still very strong, particularly in the protection of the detached house and the spacious suburb as expressions of basic prairie values. There is no doubt, however, that the prairie city is becoming infinitely more complex, more diverse, and more interesting. Much of the old monotony and uniformity is disappearing, and the new forms, whether of development or redevelopment, are designed from new social and aesthetic appreciations. There are regional variations, of course. The process has gone furthest in Edmonton and Calgary, with Regina and Saskatoon still retaining much of their pristine simiplicity. Winnipeg is in the unique situation of having a historical patina which is not shared by the other cities, and which raises special questions of conservation and preservation.

On the obverse side of the coin, the problems of the prairie cities have also grown and changed. The evident planning successes, such as the containment of urban sprawl and the organization of residential space, are matched by equally evident planning failures. The machinery for coping with redevelopment pressures, for instance, is quite inadequate, and often substitutes a new order of uniformity and monotony, at a higher density. The apartments are jammed in featureless rows and the gardens are for automobiles. Concern is also being voiced for new types of problems, such as inhuman scale, pollution, congestion, and even societal breakdown. For the first time in the prairie setting, the morality of growth is being seriously questioned. Though an acceptance of growth is still at the heart of public policy, and is probably believed in devoutly by most of the population, there is a swelling core of doubters who fear the loss of much that they cherish if the cities continue their expansionist courses. In their turn, attitudes of this kind may contribute to yet another major reshaping of the prairie city.

7   Retrospect and Prospect

J.H. RICHARDS

From southwestern Alberta the interior plains of western Canada slope gently northeastward to Hudson Bay and northward to the Mackenzie lowland. The combination of small slope and arctic drainage has been significant in the economic history of the western interior, facilitating the penetration and spread of European influence. The inference of relatively simple terrain, however, is unreal. Other elements, mainly pertaining to differences in bed-rock geology, the effects of glaciation, vegetation, climate, and soils, emerge as the determinants of major landscape regions. In particular, they help to define the complex northern forest and southern grassland sections – a separation which is palpable in terms of both Indian and European occupance, and which has been characterized by differences in type and intenstiy of exploitation.

The most acceptable term comprehending the extent and nature of the provinces of Alberta, Saskatchewan, and Manitoba is 'western interior.' It avoids the erroneous emphasis on an areally dominant grassland and suggests an important characterization vis-à-vis the rest of Canada. For, despite the late summer window of Churchill on Hudson Bay and seasonal water connections via the Mackenzie River, this is effectually a landlocked region. It is separated by barriers of distance, difficult terrain, and areas of minor economic activity from the Canadian heartland in the Great Lakes–St Lawrence lowlands and the Vancouver–Lower Mainland node. Northward of the 60th parallel, the northern boundary of the provinces, are essentially empty lands of the Northwest Territories; southward of the 49th parallel are American states with which connections are minor and in which economic development is non-complementary and of comparatively low order.

This setting suggests a general framework and constraints which have affected the history of European contacts and settlement and the forms, distribution, and intensity of economic development. It is the background against which the nature of past and present occupance may be assessed, and provides the context for interpreting and forecasting the region's changing role and status within Canada.

European Penetration and Influence: Impacts of the Fur Trade

European penetration was initiated from two directions and by two national groups. After the granting of the Charter of Rupert's Land in 1670, English fur-trading forts were established at York Factory and elsewhere on Hudson Bay. From time to time, parties were sent into the interior to encourage Indian trade and occasional contact was made with the margins of the great continental grassland. But it was not until 1774, when competition from the St Lawrence traders forced the issue, that the Hudson's Bay Company built its first inland post at Cumberland House – a good location to command traffic westward along the Saskatchewan River, northeastward to Hudson Bay, and south to the Winnipeg basin (Figure 2.1). This last had been penetrated in the 1730s by explorers and traders from New France, travelling by way of the upper Great Lakes and the Lake of the Woods. They established year-round trading posts in the Winnipeg basin, and developed the Saskatchewan route to the Athabasca country, thus threatening the fur hinterland of the Hudson's Bay Company. It was also from here that the major contact was made with Plains Indians in the trade for dried bison meat and pemmican. The Great Lakes route was not superseded until 1821, when the amalgamation of the Northwest and the Hudson's Bay companies brought the entire western interior into the hinterland of the Bay forts.

European trade induced changes in the distribution of Indian tribes, particularly in the boreal forest. By the mid-nineteenth century the relationships of the forest tribes with the fur trade were so intimate that the men became trappers tied to specific trading posts. Tribal coherence was weakened with increasing participation in the European's economy and culture. A further result was the emergence of a mixed Indian-European population – the Métis – as an important element in the economic and social structure of the fur economy. In contrast, the grasslands were peripheral to the fur grounds and the Plains Indians related to the fur economy mainly as suppliers of pemmican. With less European contact, they were able to retain their separate political organizations until the 1870s.

Until the mid-nineteenth century, economic exploitation pertained to the forests and the waterways. Contacts with the grasslands were sporadic and special, but as the character of the fur trade changed, and as Indians and Métis became increasingly dependent, a system of post settlements, river connections, and overland trails developed. The Red River colony established by Lord Selkirk in 1812 contributed much to this trend. Colonists and Métis extended settlement along the Red and Assiniboine rivers, living on strip-farms close to wood and water. The generally inefficient and disaster-ridden farming of the settlement was supplemented by highly organized bison hunts far into the grasslands. Small Métis hunting, haying, and farming settlements also sprang up, and were connected by overland trails with both the Red River settlement and the riverine trading posts. But it was still possible for Palliser to state, in 1857, that west of Fort Qu'Appelle, 'the whole country in this latitude is untravelled by the white man.'

The Western Interior: Foundation for a Canadian Dream

The Red River colony became an integral unit in the Hudson's Bay Company organization based on Fort Garry. But its location and growth accelerated the decay of the Hudson Bay route, with its seasonal dependence on polar seas. The Company was unable to maintain monopoly against local and American free traders who emphasized southward connections via the Red River. From 1849, open trade developed between the Red River settlement and St Paul, and ten years later the first steamboat came down the Red River. Thus, the physical and political separation of Rupert's Land from Canada, together with the Red River settlement's downstream location from the nearby United States, helped establish new, sometimes unwanted, commercial contacts and fears for the political security of this part of British North America.

By the 1860s, American settlement had moved into the continental interior beyond St Paul, its economic and political vitality soon expressing itself north of the 49th parallel. In contrast, the eastern territories of British North America were ill served by geography; they were separated from each other by barren lands and from the Pacific colonies of British Columbia and Vancouver Island by the great barrier of the Shield, the vast 'empty' grasslands, and the western cordillera. Only by confederation and the creation of a British North American state, through the acquisition and settlement of the interior plains, was there hope of countering the competitive commercialism and expansionist policies of the Republic. And only by establishing a separate and competitive transport system could a national integrity be maintained. Confederation came in 1867; the transfer of Rupert's Land to the Dominion of Canada in 1869; and the first transcontinental railway in 1885 with the completion of the western section of the CPR: none was an easy accomplishment.

Many surveys of the western interior were undertaken from the 1850s through the 1870s. All added greatly to the store of factual knowledge and some provided sound interpretation, not necessarily accepted, of considerable geographic significance. All were intended to assess the potential of the interior as a base for a great agricultural heartland. The dream was to recreate a context for active immigration and agricultural expansion capable of rejuvenating the commercial system of the St Lawrence, which had lost out to American frontier expansion, easier routeways, and superior transport links.

After 1867 the Dominion government inaugurated policies for the new North West Territories. These included the subdivision of the land by survey, the extinction of aboriginal claims, the allocation of blocks of land to the Hudson's Bay Company, and land grants to companies (e.g. the Canadian Pacific syndicate) participating in railway construction. Under treaties in the 1870s, Indian reservations were established on the basis of about one square mile of inalienable territory to each family of five; it was thought that through the reserve system the Indian could be transformed from hunter to sedentary farmer or labourer. Other lands were patented in the names of Métis, and yet others were set aside for sale to support schools. The remainder was available for settlement, either as free homestead lands or through sale to settlers.

Canadianization of the Western Interior:
Weaknesses in the Cement, 1870–86

The lands of the interior did not yield easily to Canada or the settler. Even before their opening to colonization in 1871, new problems were emerging. They were broadly political in nature but, as they pertained particularly to Roman Catholic, French-speaking Métis of the Red River, overtones of ethnic, religious, and social prejudice were present. The Métis feared for the security of their land titles and social order as the transfer of Rupert's Land to Canada grew imminent; American agitators, the aggressive and antagonistic attitudes of Canadian settlers, and the presence of Canada Land Survey parties did little to reassure them. Under Louis Riel they organized a provisional government in 1869, and although this was suppressed, land titles and amnesty were promised. During the period of unrest many Métis left the Red River area to take up lands elsewhere in the North-West Territories.

In 1870 the total population of the new province of Manitoba (then smaller than the modern province) was 11,963, consisting of 5757 Métis, 4083 English half-breeds, 1565 whites, and 558 Indians. Settlement was still riparian and the colony remained active in the fur trade. The natural movement of people and goods now followed the Red River and many believed that settlement of the interior would follow this route and thence westward, along traditional waterways of the fur trade. Further, their interpretation of geography stressed the park belt fringe of the grasslands as being most suitable for agricultural settlement. This evaluation was accepted by pioneers of the 1870s who welcomed in the parklands the supply of wood and water and in some cases anticipated, wrongly, that the railway would follow.

Steamboat traffic with Winnipeg was followed in 1878 by rail connection along the Red River valley, reinforcing the strong association with St Paul. After 1874, too, the steamboat appeared on the Saskatchewan River. Then, with the building of the Canadian Pacific Railway across the Shield to Winnipeg, the St Lawrence connections were to triumph, never again to be seriously challenged by either the Red River or the Hudson Bay routes. The choice of the western route of the CPR through 'Palliser's Triangle,' the heart of the dry belt, is difficult to explain, especially since the original survey projected a line through the 'fertile belt.' Macoun's interpretation of the nature of the grasslands (which he saw in a humid cycle), political considerations which suggested that the route be fairly close to the American border, and the economic concern that no hinterland products be drawn off to American railways, all influenced the decision.

Three other factors affecting settlement – the system of surveying, the early policing, and the Riel Rebellion of 1885 – are worthy of comment. The survey system of townships (each a square of six miles side) subdivided into thirty-six square sections permitted a rapid survey and the easy indentification of individual land parcels. The surveyors, too, provided notes concerning terrain, vegetation, soil, and water supply which were useful to the settler in his selection of a homestead. With few exceptions, such as some river-lot surveys, this formed the basic pattern of survey. Its impress is visible in the grid arrangement of roads, farm holdings, and municipal boundaries, and, until the recent construction of new regional highways, only the railways dared dispute the chequerboard pattern. This arrangement of space was one basis for Métis fears that their traditional and preferred long river lots would be eradicated by the survey.

The opening of the plains to settlers necessitated the formation, in 1874, of a small permanent force of North-West Mounted Police. They were to patrol the border against Indian movements following the Indian Wars in the United States and to establish Canadian law in the southwest. A series of forts was established to control the growing unrest caused by American 'whiskey traders,' by starvation among Indian bands, and by Indian resentment of white expansion, particularly as the last was interpreted in the light of the recent history of frontierism in the American plains. Bison, the staple of life, were becoming scare – the last Canadian hunt occurred in 1879 – and the reservation solution was by no means accepted by all the plains Indians. By 1884 matters were getting out of hand. Many Indians were dissatisfied with their economic plight and with the legislation formulated in distant Ottawa; some Métis were experiencing difficulty in obtaining satisfaction in their land claims; and support for Métis claims was initially given by many of the white settlers who were themselves dissatisfied with both economic conditions and government. Anti-Canadian feeling become strong and in 1885 Riel led the abortive North West Rebellion, which was quickly crushed by troops brought in by rail from the east. As in 1870, insurrection strengthened the eastern connection with the interior, while the forceful eradication of the Indian and Métis troubles cleared the way for settlement. The episode also made it apparent that the grasslands imposed few physical barriers to movement and that the old river routes were peripheral to their exploitation: railways and roads now took over.

Prairie Fabric, 1870–1900

Agricultural settlement and Canadianization were set against a physical background of remote untried lands and a social context which included nomadic Indians together with widely separated colonies of Métis and traders. During the early years immigrants were mainly from Ontario and the British Isles, but from 1874 blocks of land were reserved for groups of colonists. Reserves were established for German-speaking Mennonites from Russia, who numbered 6000 by 1879 and occupied some 60 villages in the Red River valley. French Canadians from New England and Québec began migrating in 1874, and founded 10 settlements between 1876 and 1885. Reserved blocks were given also to groups from England, Scotland, Germany, Russia, the Ukraine, Belgium, Scandinavia, and Iceland. Thus, although 'British-born' formed the dominant group, a population of strong ethnic diversity was developing, in contrast with English Ontario and French Québec.

Economic change was also apparent in the 1870s. The fur trade was still active, but optimism about railway expansion and the agricultural future of the grasslands was inducing new developments and linkages. Winnipeg was establishing itself as the gateway to the interior, developing strong commercial and industrial functions and expanding from a village of 100 in 1870 to a town of 5000 in 1875. But although population expansion and optimism were encouraging the growth of Winnipeg

Table 7.1   Number of homestead entries, western Canada, 1874–99

1874–79	1880–84	1885–89	1890–94	1894–99	Total
8,923	22,126	13,622	18,594	18,172	81,437

and of trade and manufacture, Manitoba constituted the main market for its own agricultural production. Farming was still in a pioneer, semi-commercial stage. In 1875, it showed its vulnerability to western hazards when widespread crop destruction by grasshoppers made food imports necessary. The following year was different: not only did the grain crop provide a surplus, but a small shipment of Red Fife seed grain was sent to Toronto.

As a consequence of a number of factors which were roughly contemporaneous in their appearance, a new economic base was being established in the late 1870s and some of the expectations of Confederation apparently were to be realized. Red Fife, introduced as a replacement seed following crop failure, was far superior to wheat previously grown and quickly became the dominant commercial type. With excellent milling and baking qualities and a maturation period (115 to 125 days) 10 to 20 days shorter than wheat strains commonly grown in the Red River area, it was well adapted to the short growing season, to the export market, and to the new roller milling process. From 1877 onward, American and eastern buyers entered the grain market. At the same time railway building became active, work on the branch to Pembina to connect with the American line commencing in 1875 and on the section of the Canadian Pacific eastward from Selkirk in 1876. By 1879 Mennonite and other settlers were demonstrating that the open plains could support general farming. The extension of settlement away from river and woodland was aided by the availability of agricultural equipment tested in the American Midwest, including the steel windmill, barbed wire, chilled steel mouldboard plow, and self-binding reaper. Within the next 10 or 15 years black fallow, to conserve moisture so as to produce a crop the following year, was incorporated into a dry farming system which made possible the agricultural occupation of the dry plains.

Movement of the wheat culture on to the open plains was relatively slow and sporadic (Table 7.1), sensitive alike to economic conditions in other countries and to climatic and other hazards of the grasslands. The first 30 years of settlement also established distribution patterns of population, agricultural land use, and resource development which have persisted to the present (Figure 5.1). At the end of the century, settlement was still concentrated in southern Manitoba. It had spread along the transcontinental railway as far as Moose Jaw, but there were very few settlers on the semi-arid lands west of the Missouri Côteau. Similarly, there was a pronounced zone of settlement through the parkland of eastern Saskatchewan, but west from Saskatchewan forks it was spread much more thinly along the river system. North-reaching branch railways from Regina to Saskatoon and Prince Albert, and from Calgary to Edmonton, provided further axes for settlement. In effect, the settlers' preference for the humid margins of the grasslands had been declared, though, in part, this distribution reflected both historical accident and the linkages afforded by railway and river. The Saskatchewan system, particularly along its north branch, was an active transport route for three decades, though its importance diminished with each new railway connection.

Among other resources, natural gas was in use as a fuel in southwestern Alberta, and coal mines between Edmonton and Lethbridge and in the adjacent Rockies supplied the railways and markets as far east as Winnipeg. Lignite deposits were being worked in southeastern Saskatchewan, brickclays were extracted in all three provinces, and salt was mined mainly in Manitoba. There were also lumber from the eastern slopes of the Rockies, from Manitoba, and from elsewhere on the forest fringe; fish, mainly from the Manitoba lakes; and a promise of minerals and water power from the Shield. But major developments of these were to wait on the next century: more important were the national and international relationships now being established by wheat production and export.

Already regional differences in agriculture were apparent. Manitoba dominated wheat production, but it had also developed considerably in mixed farming. Its advantages were a climate somewhat more humid than farther west, a relatively strong local market, and, after 1889, the opportunity to ship live cattle to eastern Canada and Great Britain. Westward as far as Moose Jaw wheat was the cash crop, although other crops were grown for forage and most farms tended to be self-sufficient in food production. Because of climatic hazards, diversification was encouraged, not too successfully, and experimental farms were set up. In southern Alberta and southwestern Saskatchewan, ranching was well entrenched but grain farming was moving in. Irrigation of fodder and other crops was of growing importance in southwestern Alberta so that by 1895 some 80,000 acres were irrigated in over 100 projects. Droughts in the 1880s and 1890s caused much concern about the validity of agricultural settlement in the western interior; they led to the

Table 7.2   Number and capacity (millions of hectalitres) of grain elevators (including warehouses) to 1910

Year	Manitoba		Saskatchewan		Alberta	
	Number	Capacity	Number	Capacity	Number	Capacity
1879	1	–	–	–	–	–
1890	128	1.4	65	0.1	–	–
1900	469	4.7	93	0.7	18	0.1
1910	719	7.9	909	9.7	262	3.2

North-West Irrigation Act of 1894 which provided the basis for controlling the future allocation and use of water in the prairie provinces. By 1900, the Edmonton area was joining in the wheat-growing venture, but mixed farming was more typical, as it was for the Saskatchewan forks region. Near Saskatoon, ranching still was common.

Table 7.2 indicates the progression westward of commercial wheat farming but it does not show that Saskatchewan's 65 elevators in 1890 were located at only 16 points, Regina being the farthest west. It reflects the great growth in wheat production to the end of the century and, by inference, suggests the impact of the wheat economy on regional investment in elevators and railways, and its concomitant effects upon the national economy in similar and supportive enterprises ranging from port facilities to manufacturing and wholesaling. The first 30 years of western settlement had been hesitant and experimental but, in total, fairly successful. Much of the pattern for development, in terms of area and type, had been established as had a history of in-migration and out-migration which was to have its parallels in the twentieth century. Population influx followed by loss of part of the gain is not atypical of the Canadian experience; it had been – and was to be – especially characteristic of the prairie region, where primary production was susceptible to both natural and market hazards.

The Prairie Provinces in the Twentieth Century

The prairie region was both population magnet and major economic stimulant during the first quarter of the twentieth century. Its population increased from 491,512 in 1901 to 2,353,524 in 1931, an increase involving mainly rural population (63 per cent of the total in 1931) but initiating considerable urban growth (Table 4.1) and local manufacturing activity; capital investment in the last rose from 9 million dollars in 1901 to 84 million dollars 10 years later. Of the 25,000 miles of railways in Canada in 1910, nearly one-third were in the prairie prov-

Table 7.3   Prairie farm and transport investment, 1901–30
(in millions of dollars)

	Canadian gross capital formation (A)	Prairie farm investment (B)	B as % of A	Transport investment (C)	C as % of A
1901–05	1284	221	17.2	201	15.7
1906–10	2241	319	14.2	539	24.1
1911–15	3230	463	14.3	848	26.3
1916–20	4138	370	8.9	661	16.0
1921–25	3702	245	6.6	753	20.3
1926–30	5898	454	7.7	1225	20.8

Source: Data from K.E.A. Buckley, *Capital Formation in Canada 1896–1930* (University of Toronto Press, Toronto, 1955).

inces, and they accounted for half of the 11,000 miles built during the next five years. In motor vehicle registrations, the region accounted for more than a quarter of the Canadian total in 1910 and for one-third in 1920. Between 1901 and 1931, 38 million hectares of farmland were added to the 6 million occupied by 1901, the area in field crops alone increasing from 1.4 million to 16 million hectares. This highly successful occupation of the interior and the burgeoning wheat economy induced the first considerable economic interaction between the regional separates of Canada.

By 1918 wheat had become Canada's most important single export in dollar value and the prairie provinces loomed large in the context of Canadian economic growth. The influence of prairie farm and transport investment on gross Canadian capital formation is portrayed in Table 7.3 as is the large capital investment necessary to develop the region. What is not shown is that western grain farmers competed in an international market but purchased manufactured products at prices which were affected by a tariff structure designed to protect and encourage manufacturing in central Canada. Indeed, then and now, the cry of the paririe provinces has been for freer trade in manufactured items and lower transportation costs, so as to reduce the selling costs of prairie products and the buying costs of goods manufactured elsewhere.

A peak of wheat production, in both quantity and quality, came in 1928. It was followed by the period of general economic depression coinciding with a decade of natural disaster during which drought, wind erosion, and grasshopper infestation devastated the core of the Palliser Triangle, and extended over most of the cropped area at some time.

Improved weather conditions and the demands of the war and post-war years helped re-establish confidence, and the area in grain rebounded, by 1966, to 19 million hectares, including some 12 million in wheat. In 1968 a record wheat crop was harvested, but by the last years of the sixties a glut of grain on the world market commenced a new era in the western adventure: a system of controls and bonuses was put into effect to divert wheat land into other forms of production, which is difficult in the prairie climate, or into retirement. The grain farmer had not overcome the environmental and market hazards, but he had succeeded only too well in accommodating his specialized, highly mechanized farming system to them. The status of wheat in the national and regional economies has deteriorated greatly, but as late as 1966 wheat exports, excluding flour, constituted one-tenth of the total value of domestic exports from Canada. To wheat, barley, oats, rye, flaxseed, and rapeseed must be added other agricultural production, not the least of which is a very considerable cattle population not kept for milk (i.e., mainly beef cattle), amounting to over six million or two-thirds of the Canadian total.

The one-fifth of the Canadian area occupied by the prairie provinces supported 23 per cent of the Canadian population in 1931. This population focused on the grasslands and farming, its greatest densities coinciding with the humid margins. Northward were pockets of farm settlement occupying grassland exclaves; the largest, the Peace River District, was to achieve considerable growth during the exodus from drought areas in the 1930s and later. The forested area, peripheral and tributary, still yielded furs, but lumbering and pulpwood production came to have greater importance after the Pine Falls, Manitoba, paper mill was built in the late 1920s. On the Shield major metallic mineral finds had been made, particularly at Flin Flon (Figure 4.1). This complex of copper, zinc, gold, silver, and other minerals had to wait until 1930 for hydroelectric power and rail transportation to the south; it was also linked with the Hudson Bay Railway, which re-established a northern route to the port of Churchill in 1931. Such relationships between the grassland and the north were to increase in strength.

Until 1931 agriculture provided the support for settlement, but as the rural population declined to 37 per cent of the regional total in 1966, the prairie provinces' share of the national population fell to 17 per cent (Table 5.1 and Figure 5.2). Only in Saskatchewan is agriculture still the leading industry by value added (Tables 1.1 and 3.3); it has also remained the most rural of the provinces and has shown the least population growth since 1951 (Table 5.4). The proportional decline of the region would have been much greater had it not been for new-found strengths in the economic base, more particularly in Alberta where new contacts with the northwest were given impetus by military requirements during World War II, and where major petroleum and natural gas discoveries have followed the bringing in of the Leduc field in 1947. As of 1971, Alberta became Canada's leading province in dollar value of mineral production ($1650 million). It produced over 70 per cent of Canada's petroleum output, the bulk of its natural gas, and much of its coal; all three items are significant international exports. The fuel industry has helped to re-emphasize the long axial development from the northwest through Edmonton and Calgary to the United States.

Smaller growth of fossil fuel mining has been typical of Saskatchewan, which accounts for about one-fifth of Canada's petroleum production, together with a much smaller quantity of natural gas. Lignite is also mined for thermal electricity. Potash mines, sunk in the 1960s, are capable of extracting 12 million tons annually, but market conditions are such that only some 3 million tons are produced – enough to make the province the world's leading potash exporter. Other products include sodium sulphate, helium, uranium, and base metals from the Shield. Manitoba's major mineral resources come from the Precambrian area and include some large ore bodies of nickel, copper, and zinc; considerable hydroelectric power is also derived from rivers of the Shield. The contribution of the forest to the prairie economy is far less than that of minerals, but lumber, pulp, and paper mills and other forest industries are found in all provinces. Alberta's forests are the most extensive and productive (Table 1.1).

Participation in manufacturing is meagre, representing less than 8 per cent of the total value of shipments of goods of Canadian manufacture in 1968. Alberta accounts for one-half and Manitoba for one-third of prairie shipments. Major types of manufacturing are food (including slaughtering and meat packing) and beverages, petroleum processing, metal fabrication, primary metal processing, and machinery. Winnipeg leads in manufacturing activity, equalling the combined total of Edmonton and Calgary; these cities are also the chief wholesale centres for the prairie provinces.

Some regional differences in the type of and relative dependence upon agriculture and other resources have been suggested. These, in part, are reflected in the size and activities of prairie cities, but neither regional differences in land use nor location on the humid margins are sufficient to explain the increasing dominance of Winnipeg, Edmonton, and Calgary. This explanation relates to their history and kinds of

Table 7.4   Projected growth of total population, prairie provinces and Canada (in thousands, with the percentage of the Canadian population in parentheses)

Year		Canada	Manitoba	Saskatche-wan	Alberta
1981	High projection	27,129	1124 (4.2)	1054 (3.9)	2124 (7.8)
	Low projection	24,858	1018 (4.1)	966 (3.9)	1801 (7.3)
2001	High projection	41,568	1372 (3.3)	1107 (2.7)	3537 (8.5)
	Low projection	30,345	960 (3.2)	854 (2.8)	2121 (7.0)

Source: Systems Research Group, *Canada: Population Projections to the Year 2000* (Toronto, 1970), as quoted in J.W. MacNeill, *Environmental Management* (Information Canada, Ottawa, 1971), 188.

development together with changes in their functional linkages, the last affected by the building of new regional highways, dieselization of railways, and new airway and pipeline connections. Saskatoon and Regina are centrally located within the prairie oecumene and were appraised in the 1950s as having very considerable growth potential: they are now entering a difficult stage when economic concentration (i.e., centralization, and not centrality) has become effective so that industries such as oil refining, flour milling, and meat packing and certain service functions have either been moved or are threatened with removal to the larger prairie cities (to Edmonton, for example, for oil refining) or even to American centres (as in the case of some farm implement operations). Contemporaneously, Vancouver has contrived not only to extend its wheat export umland farther into the interior so as to threaten the traditional Lakehead supremacy but also to enlarge its service operations so as to affect the whole prairie region. And the metropolitan dominance of central Canada, so long exerted but now rendered easier by modern communications, continues unabated despite the rise of the three large prairie centres. In Table 7.4 it is suggested that, although absolute increase in population may be expected for the region, it is likely that provincial differences will be greatly accentuated, with Alberta entrenching its dominant role. These projections also direct attention to a significantly declining status (peripherality?) of the prairie region with respect to Canada.

Prairieland: Conclusion

Eastern Canadian attitudes toward the western interior, more specifically toward 'prairieland,' have been changeable but only occasionally approving, as during the brief period of the first quarter of this century when it was accepted as a land of opportunity. Mostly it has been viewed in terms of remoteness and isolation, as being vaguely – or definitely – inferior in cultural and economic status, and (according to one modern geographer) of presenting 'pressing ethnic problems.' Even the westerner might accept 'remoteness and isolation' as not an unreasonable characterization, and he is likely to accept that the region is peripheral to the areas of major economic activity in Canada. Other assessments are more difficult to accept or authenticate. Certainly, the opening of the west was not costless for Canada, and the disastrous years of the 1930s necessitated massive relief programmes, in Saskatchewan especially. Today, in the early 1970s, the costs of equalization payments made through the federal government to 'have-not' provinces, one of which is Saskatchewan, are being questioned by some of the economically stronger provinces; and, similarly, there is recognition of subsidy for the grain farmer in the recent introduction of a two-price (domestic and export) system for wheat. Against this appraisal of second-class economic status there are good arguments. The opening of the west not only made, and makes, Canada possible but it created the staple trade upon which modern Canada was built; and it has continued to supply raw materials, including grain, which have been basic not only to Canadian export trade but also to the continued development of the metropolitan east and the Pacific focus of the Vancouver–Lower Mainland region.

Possibly the concepts of ethnic problems and 'inferior cultural status' are related and may be so discussed. It took approximately 2 million incomers to establish the 1931 population of 2.3 million. These included people of diverse origins – Canadians, Britons, Americans, Ukrainians, Poles, Russian-Germans, Scandinavians, Dutch, Austrians, Lithuanians, Bohemians, and Hungarians. The variety of languages and cultures created some difficulties, which were perpetuated by the fact that many non-English-speaking groups established themselves so as to create culturally homogeneous blocs within the farmed areas. But language and other barriers stemming from this were soon partially or wholly breached by the imposition and acceptance of a common school system and common British institutions. Indeed, the process of prairie settlement had been wasteful: only a residual of 800,000 of the original foreign-born immigrants remained in 1931, the bulk having moved elsewhere. Thus by 1931, with nearly half its people 'prairie born,' the population was becoming indigenous; following the end of World War II, ethnic characterization and discrimination disappeared and (except for Indians and Métis, and the small religious colonies) acceptance in all activities has been easy and equal. This is not to suggest that strong cultural differences do not exist; they were shown, for example, in the response to the Commission on Bilingualism and Biculturalism when prairie groups made a case for the inclusion of Ukrainian as one of the official languages of Canada. In no way, however, is there a 'pressing ethnic problem' and in no way could one be demonstrated in voting or other patterns.

If there is an outsider's assessment of the prairies which tends to be less than enthusiastic there is also a prairie viewpoint which may be less than kind to Canada. Best expressed by Morton (1967), this appreciation is shared, if not enunciated as succinctly, by many westerners: 'Confederation was brought about to increase the wealth of Central Canada, and until that original purpose is altered, and the concentration of wealth and population by national policy in Central Canada ceases, Confederation must remain an instrument of injustice.' In another essay (Morton 1955) he discusses the bias of prairie politics which 'began in colonial subordination, continued in agrarian revolt and went to political and economic utopianism.' Professor Morton recognizes western regionalism and sectionalism, deriving them both from and directing them at a Canada which has been less than perfect in her dealings with the prairie provinces. The reality of the prairie region is not to be doubted even though it may have a basis in a sense of regional disparity and colonial subordination, as expressed in a recent statement of a prairie provincial premier: 'The problems of the West are economic ... the constant catering to the vested industrial interests of Central Canada ... centred around the "Golden Horseshoe" in southern Ontario.'

It is easy to trace, if not to accept, the historical bases upon which Morton predicates his interpretation of the bias of prairie politics. With others he defines the prairie west as a colonial society seeking equality in Canada, and inherent in such assessment is the recognition of a regional entity. The focus of this region is the grassland interior formerly dominated by the grain economy but including, typically, other primary production for export. Although it may be argued that the western interior does not constitute a single functional region and that its parts exihibit considerable physical and economic differences, that many of its economic linkages have changed or are in process of change, it would be more difficult to suggest that the differences and changes are of such magnitude that they vitiate the concepts of closed interior and peripherality, of difficult climate and related difficulties in agricultural adaptations, and of continuing involvement in the production of primary products for foreign or distant domestic markets.

Climatic and related difficulties have been exposed variously at different times and in different areas, the zonal differences in farm types representing areal adaptations which have developed after considerable experimentation and planning. This agricultural sector of the economy has been both bane and balm to the prairie people, who have contended with crop disaster and plenty, with bad and good markets, and, even, with inadequate or good market procedures and facilities. In doing so, they developed in common a certain willingness to experiment politically, moving into cooperative systems and accepting new ideologies; they have been willing to accept, also, direct government help and intervention. The prairie producer has consistently recognized that external markets are beyond his control, but his attack on monopoly and inadequate grain storage, handling, and market facilities in the late nineteenth century led, in 1899, to the first Royal Commission to enquire into the grain trade. Later, and with considerable provincial government assistance, the farmers moved into cooperative grain marketing and elevator ownership, marketing eventually being taken over by the federal government's Wheat Board. Other cooperative developments in retailing, wholesaling, and manufacturing (animal feedstuffs and fertilizer, for example) followed.

Direct intervention by governments into the farm economy have been necessary, and generally willingly accepted. The effects of drought and wind erosion in the 1930s necessitated massive intervention when the provincial governments and the government of Canada, through its Prairie Farm Rehabilitation Administration, removed vast areas of light soils from cultivation, introduced new techniques in farming, brought in new irrigation schemes, and established a system of stock watering dugouts. More recently, because of glutted markets, federal intervention has taken the form of subsidies for removing land from cultivation and the strong encouragement of diversification in agriculture – the last a difficult result to achieve, especially within the Palliser Triangle.

Primary producers of bulk products recognize both the lack of a large prairie market and the high costs of transport to external markets. Natural gas and petroleum proceed to Ontario and the United States via pipelines but other products – grain, coal, pulp and paper, potash – move mainly by rail. Rail transport costs have been a concern since the 1880s when attempts were made to obtain a line to Hudson Bay to break the CPR monopoly, and similar attempts to reduce freight costs have been made since, by all industries. Today, for instance, the cost of rail transport prevents the entry of Alberta coal into the central Canadian market. Concern is still expressed over the small use made of the port of Churchill. It is argued that the navigation season is considerably longer than that now permitted and that the port, which brings Liverpool 2000 kilometres nearer to the prairie provinces, could provide significant savings if given intensive and extended use. Blame for not further developing the port is awarded 'eastern Canadian interests': without discussing whether such blame is proper, the fact that little is imported through Churchill is a measure of the economic strength of the prairie region. Of common concern, too, is the abandonment and proposed abandonment of thousands of miles of branch railways. Farmers are forced to haul grain longer distances by truck, and the decrease in the number of branch elevators induces a further reduction of the few functions of many villages. Probably the greatest single concern is the movement of grain to ports during times of great demand, when for various reasons, including slides in the mountains and poor processing procedures at the docks, there have been costly delays at Vancouver. Similar difficulties have been encountered in exporting Alberta coal to Japan. The concepts of a 'western interior' and a 'colonial economy,' then, are not inadequate.

Neither is the concept of underdevelopment ill-applied. In Saskatchewan and in Manitoba, particularly, the wish to bring about diversification in primary production and in manufacturing has been such that considerable government investment capital or guarantees have been given to enterprises in which private investment is of a relatively low order. In all three provinces it would seem that environmental quality – and very real damage – may be ignored in favour of industry. Such eagerness to attract capital and industry is typical of depressed areas and peripheral regions characterized by small diversity and low participation in manufacturing. Yet it should not be suggested that this is a poor region, though it may be so at times; since 1946 the personal income per capita has approximated the national average and as late as 1966, at $2172, it was slightly above (Canada, $2134; Alberta, $2223; Saskatchewan, $2235; Manitoba, $2044).

It is rather strange that Saskatchewan, except between 1964 and 1971, has had a long term of socialist government which has shown little interest in regional planning; yet, with its present arrangement of hundreds of rural municipalities, cities, and towns, and a hodge-podge of sector 'planning and administrative' areas, it presents a seemingly chaotic approach to resource and functional planning. This does not infer that sector planning has been absent: in education, health, agriculture, social services, and recreation, for example, excellent work and standards have been achieved. Alberta has created a number of physical planning regions based on cities or major towns, but has not yet considered forms of regional government or economic planning. In Manitoba, a system of 'economic regions' was created in the 1950s to assist in economic planning. A 1963 report on Manitoba's future to 1980 is interesting in that it attempts to deal with the problem of scattered rural and small village population; its proposals included reducing the number of farms to some 20,000, increasing net income per farm by one-third, and concentrating population into the towns. Actually, these phenomena are occurring naturally, if slowly, across the region. It is probably of advantage to ensure, as far as possible, that the process is orderly and worth while in all its social and economic implications. In doing so, in selected areas such as the Inter-Lake region of Manitoba, full use is being made of federal aid through special programmes relating to regional disparities. Such palliatives are piecemeal, however, and are not related to any regional policy of population relocation. Meanwhile, the drift to the cities continues unabated and apparently unstoppable.

The reality of the prairie region has given rise, on occasion, to some concern for closer formal relationships. This is best expressed in the Prairie Provinces Water Board, which was set up in 1948 'in the light of the scarcity of water resources in the prairies, to determine and recommend the best use to be made of interprovincial waters ... and to recommend the allocation as between provinces.' It is to be noted that water scarcity is a characteristic of the southern parts of the provinces, not of the north. Common interests also gave rise to a Prairie Provinces Economic Council, of no great apparent strength. More recently the concept of a single prairie province has been discussed publicly, but provincial differences, particularly in economic stature, are sufficient to inhibit formal organization.

Despite the fact that provincial boundaries are, de facto, both political and regional separators, the reality of the prairie region – the western interior – is undeniable. Whether fighting 'eastern interests' and Ottawa, threatening secession (as in the 1930s), 'counterbalancing' the demands of new Québec, staggering from overproduction or underproduction of grain, suffering recession or being different in politics, the parts are one. Recognizing a gap of two generations in the time of its development and the nature of its present economy, the region, as always in frontierism, rejects eastern opinion in favour of its own interpretation of Canada.

References

Abramson, J.A., 1968 *Rural to Urban Adjustment* (Department of Forestry and Rural Development, Ottawa)

Anderson, J., 1968 Economic Base Measurement and Changes in the Base of Metropolitan Edmonton, *The Albertan Geographer*, 4: 4–9

Baker, W.B., 1958 Changing Community Patterns in Saskatchewan, *Canadian Geographical Journal*, 56 (2): 44–56

Banfield, A.W.F., 1958 *Mammals of Banff National Park* (Bulletin 159, Canada Department of Northern Affairs and National Resources, Ottawa)

Bennett, J.W., 1969 *Northern Plainsmen* (Aldine Publishing Co., Chicago)

Bird, R.D., 1961 *Ecology of the Aspen Parkland of Western Canada* (Publication 1066, Canada Department of Agriculture, Ottawa)

Brown, R.J.E., 1969 *Permafrost in Canada* (National Research Council of Canada, Publication no. NRC 9769, Ottawa)

Budd, A.C., 1957 *Wild Plants of the Canadian Prairies* (Experimental Farms Service, Publication 983, Canada Department of Agriculture, Ottawa)

Burpee, L.J., 1907 York Factory to the Blackfeet Country: The Journal of Anthony Henday, 1754–55, *Proceedings and Transactions of the Royal Society of Canada*, series 3, section II: 307–61

Byrne, A. Roger, 1968 *Man and Landscape Change in the Banff National Park Area before 1911* (Studies in Land Use History and Landscape Change, National Park Series, no. 1, University of Calgary)

Canada, Department of Agriculture, 1970 *The System of Soil Classification for Canada* (Department of Agriculture, Ottawa)

Canadian National Committee for the International Hydrologic Decade, 1969 Preliminary Maps, Hydrological Atlas of Canada (Canadian National Committee, Ottawa)

Chapman L.J., and D.M. Brown, 1966 *The Climates of Canada for Agriculture* (The Canada Land Inventory, Report no. 3, Department of Forestry and Rural Development, Ottawa)

Coues, E., 1965 *New Light on the Early History of the Greater Northwest: The Manuscript Journals of Alexander Henry and of David Thompson, 1799–1814* (Ross and Haines, Minneapolis)

Craddock, W.J., 1970 *Interregional Competition in Canadian Cereal Production* (Economic Council of Canada, Special Study no. 12, Ottawa)

Crowley, John M., 1961–2 'Geographic Aspects of the Canadian Oil Industry,' *Cahiers de Géographie de Québec*, 11: 97–107

Dawson, J., 1965 *Changes in Agriculture to 1970* (Economic Council of Canada, Staff Study no. 11, Ottawa)

Douglas, R.J.W. (ed.), 1970 *Geology and Economic Minerals of Canada* (Department of Energy, Mines and Resources, Economic Geology Report no. 1, Ottawa)

Dunlop, J.S., 1970 Changes in the Canadian Wheat Belt, 1931–1969, *Geography*, 55 (2): 156–68

Ehlers, Eckart, 1966 The Expansion of Settlement in Canada: A Contribution to the Discussion of the American Frontier, *Geographische Rundschau* 18 (9): 327–37

Elton, D.K. (ed.), 1970   *Proceedings of One Prairie Province? A Question for Canada and Selected Papers* (Lethbridge Herald, Lethbridge)

Fidler, P.   *Journal of a Journey over Land from Buckingham House to the Rocky Mountains, in 1792–93* (Hudson's Bay Company Documents, E 3/2, Reel 4M4, Public Archives of Canada, Ottawa)

Fuller, W.A., 1961   Emerging Problems in Wildlife Management, *Resources for Tomorrow: Conference Background Papers*, II (Ottawa): 884–8

Furniss, I.F., 1968   Productivity Trends in Canadian Agriculture, 1935 to 1965, in R.M. Irving (ed.), *Readings in Canadian Geography* (Holt, Rinehart and Winston, Toronto/Montreal): 205–15

Gagan, D.P. (ed.), 1970   *Prairie Perspectives* (Holt, Rinehart and Winston, Toronto/Montreal)

George, M.V., 1970   *Internal Migration in Canada* (Dominion Bureau of Statistics, 1961 Census Monograph, Ottawa)

Gertler, Leonard O., 1960   Some Economic and Social Influences on Regional Planning in Alberta, *Plan Canada*, 1 (2): 115–21

Glover, R. (ed.), 1962   *David Thompson's Narrative, 1784–1812* (Champlain Society, Toronto)

Government of Alberta and the University of Alberta, 1969   *Atlas of Alberta* (University of Alberta Press, Edmonton)

Gray, Earle, 1969   *Impact of Oil* (Ryerson Press/Maclean Hunter, Toronto)

Gray, James H., 1967   *Men Against the Desert* (Modern Press, Saskatoon)

Hansen, H.P., 1949   Post-glacial Forests in South-Central Alberta, *American Journal of Botany*, 36: 54–65

Hanson, Eric J., 1958   *Dynamic Decade: The Evolution and Effects of the Oil Industry in Alberta* (McClelland and Stewart, Toronto)

Hardy, W.G. (ed.), 1967   *Alberta: A Natural History* (M.G. Hurtig, Edmonton)

Hodge, Gerald F., 1965   The Prediction of Trade Center Viability in the Great Plains, *Papers of the Regional Science Association*, 5: 87–115

— 1966   Urban Systems and Regional Policy, *Canadian Public Administration*, 9 (2): 181–93

— 1968   Branch Line Abandonment: Death Knell for Prairie Towns? *Canadian Journal of Agricultural Economics*, 16 (1): 54–70

Innis, H.A., 1964   *The Fur Trade in Canada* (University of Toronto Press, Toronto)

Larson, F., 1940   The Role of Bison in Maintaining the Short Grass Plains, *Ecology*, 21 (2): 113–21

Laycock, A.H., 1967a   A Present Land Use Map of Alberta, *The Albertan Geographer*, 3: 35–40

— 1967b   *Water Deficiency and Surplus Patterns in the Prairie Provinces* (Prairie Provinces Water Board, Report no 13, Regina)

Lenz, Karl, 1963   Die Gross-Städte im mittleren Western Kanadas; ihre Entwicklung und Stellung innerhalb der Provinzen, *Geographische Zeitschrift*, 51 (4): 301–23

— 1965   *Die Prärieprovinzen Kanadas* (Marburger Geographische Schriften, Heft 21, Marburg)

Longley, R.W., 1967   'The Frequency of Chinooks in Alberta,' *The Albertan Geographer*, 3: 20–2

Longley, R.W., and J.M. Powell, 1970   *Bibliography of the Climate of the Prairie Provinces, 1957–1969* (University of Alberta Press, Edmonton)

Lupton, A.A., 1967   Cattle Ranching in Alberta, 1874–1910: Its Evolution and Migration, *The Albertan Geographer*, 3: 48–58

Lycan, Richard, 1969   Interprovincial Migration in Canada: The Role of Spatial and Economic Factors, *The Canadian Geographer*, 13 (3): 237–54

Mackintosh, W.A., and W.L.G. Joerg (eds.), 1934   *Prairie Settlement: The Geographical Setting*, vol. 1 in *Canadian Frontiers of Settlement* (Macmillan Company of Canada, Toronto)

McCann, L.D., 1969   Urban Growth in Western Canada, 1881–1961, *The Albertan Geographer*, 5: 65–74

McCrossan, R.G., and R.P. Glaister, 1966   *Geological History of Western Canada*, 2nd ed. (Alberta Society of Petroleum Geologists, Calgary)

McDougall, J., 1902   *George Milliard McDougall: Pioneer, Patriot and Missionary* (W. Briggs, Toronto)

— 1911   *Our Western Trails in the Early Seventies: Frontier Pioneer Life in the Canadian North-west* (W. Briggs, Toronto)

McDougall, J.L., 1968   *Canadian Pacific: A Brief History* (McGill University Press, Montreal)

McIntosh, R. Gordon, and Ian E. Housego (eds.), 1970   *Urbanization and Urban Life in Alberta* (Alberta Human Resources Research Council, Edmonton)

Morton, W.L., 1955   The Bias of Prairie Politics, *Transactions of the Royal Society of Canada*, series 3, 49, section 2: 57–66

— 1967   Clio in Canada: The Interpretation of Canadian History, in Carl Bergen (ed.), *The Interpretation of Canadian History*, (Canadian Historical Readings no. 1, Toronto)

Moss, E.H., 1955   The Vegetation of Alberta, *Botanical Review*, 21: 493–567

Nader, G.A., 1971   Some Aspects of the Recent Growth and Distribution of Apartments in the Prairie Metropolitan Areas, *The Canadian Geographer*, 15 (4): 307–17

Neilsen, K.F. (ed.), 1967   *Proceedings Canadian Centennial Wheat Symposium, Saskatoon, 1967* (Western Cooperative Fertilizers Ltd., Saskatoon)

Nelson, J.G., 1970   Man and Landscape Change in Banff National Park: A National Park Problem in Perspective, in J.G. Nelson (ed.), *Canadian Parks in Perspective* (Harvest House, Montreal): 63–96

Nelson, J.G., and M.J. Chambers (eds.), 1969–70 *Process and Method in Canadian Geography* (4 vols., Methuen, Toronto)

Nelson, J.G., and R.F. England, 1971 Some Comments on the Causes and Effects of Fire in the Northern Grasslands Area of Canada and the Nearby United States, ca. 1750–1900, *The Canadian Geographer*, 15 (4): 295–306

Peet, J. Richard, 1963 Natural Gas Industries in Western Canada, *The Canadian Geographer*, 7 (1): 23–32

Raby, Stewart, 1964 Alberta and the Prairie Provinces Water Board, *The Canadian Geographer*, 8 (2): 85–91

— 1965 Irrigation Development in Alberta, *The Canadian Geographer*, 9 (1): 31–40

Richards, J.H., 1967 A Discussion of Problems in Resource Development: Overtones of Human Frailty, *The Albertan Geographer*, 3: 23–8

— 1968 The Prairie Region, in John Warkentin (ed.), *Canada: A Geographical Interpretation* (Methuen, Toronto): 396–437

Richards, J.H., et al., 1969 *Atlas of Saskatchewan* (University of Saskatchewan, Saskatoon)

Roe, F.G., 1970 *The North American Buffalo: A Critical Study of the Species in Its Wild State*, 2nd ed. (University of Toronto Press, Toronto)

Ross, Eric, 1970 *Beyond the River and the Bay: Some Observations on the State of the Canadian Northwest in 1811* (University of Toronto Press, Toronto)

Ruggles, Richard I., 1971 The West of Canada in 1763: Imagination and Reality, *The Canadian Geographer*, 15 (4): 235–61

Scace, Robert C., 1968 *Banff: A Cultural-Historical Study of Land Use and Management in a National Park Community to 1945* (Studies in Land Use History and Landscape Change, National Park Series, no. 2, University of Calgary)

Smith, P.J., 1962a Fort Saskatchewan: An Industrial Satellite of Edmonton, *Plan Canada*, 3 (1): 4–16

— 1962b 'Calgary: A Study in Urban Pattern,' *Economic Geography*, 38 (4): 315–39

— 1971 Change in a Youthful City: The Case of Calgary, Alberta, *Geography*, 56 (1): 1–14

Stone, Leroy O., 1967 *Urban Development in Canada* (Dominion Bureau of Statistics, 1961 Census Monograph, Ottawa)

Stutt, R.A., 1971 Changes in Land Use and Farm Organization in the Prairie Area of Saskatchewan during the Period 1951 to 1961, *Canadian Farm Economics*, 5 (6): 11–19

Swainson, Donald (ed.), 1970 *Historical Essays on the Prairie Provinces* (McClelland and Stewart, Toronto/Montreal)

Szabo, M.L., 1965 Depopulation of Farms in Relation to the Economic Conditions of Agriculture on the Canadian Prairies, *Geographical Bulletin* 7 (3–4): 187–203

Vanderhill, B.G., 1962   The Decline of Land Settlement in Manitoba and Saskatchewan, *Economic Geography*, 38 (3): 270–7

— 1963   Trends in the Peace River Country, *The Canadian Geographer*, 7 (1): 33–41

Van Vliet, H., 1961   The Prairie Agricultural Region, *Resources for Tomorrow: Conference Background Papers*, vol. 1 (Ottawa): 527–37

Warkentin, John, 1963–4   Western Canada in 1886, *Papers*, Historical and Scientific Society of Manitoba, series III, 20: 85–116

— (ed.), 1964   *The Western Interior of Canada: A Record of Geographical Discovery 1612–1917* (McClelland and Stewart, Toronto/Montreal)

— (ed.), 1968   *Canada: A Geographical Interpretation* (Methuen, Toronto)

Watts, F.B., 1960   The Natural Vegetation of the Southern Great Plains of Canada, *Geographical Bulletin*, 14: 25–43

Weir, Thomas R., 1960   *Economic Atlas of Manitoba* (Department of Industry and Commerce, Winnipeg)

— 1964a   Pioneer Settlement of Southwest Manitoba, 1879 to 1901, *The Canadian Geographer*, 8 (2): 64–71

— 1964b   *Rural Population Change and Migration: Interlake District, Manitoba* (Manitoba Department of Agriculture and Conservation, Winnipeg)

Weir, Thomas R., and G. Matthews, 1971   *Atlas of the Prairie Provinces* (Oxford University Press, Toronto)

Wonders, William C., 1959   River Valley City–Edmonton on the North Saskatchewan, *The Canadian Geographer*, 14: 8–16

Wormington, A.M., and R.G. Forbis, 1965   *An Introduction to the Archeology of Alberta, Canada* (Denver Museum of Natural History, Proceedings, no. 11, Denver)